CATALYSIS

Progress in Research

CATALYSIS
Progress in Research

Proceedings of the NATO Science Committee Conference on Catalysis
held at Santa Margherita di Pula, December 1972

Edited by
Fred Basolo
and
Robert L Burwell, Jr.

Northwestern University, Department of Chemistry
Evanston, Illinois, 60201, *U.S.A.*

Published in coordination with
NATO Scientific Affairs Division by
PLENUM PRESS • LONDON and NEW YORK • 1973

Library of Congress Catalog Card Number 73-81490
ISBN 0-306-30753-7

Plenum Publishing Company Ltd.
Davis House, 8 Scrubs Lane, Harlesden
London NW10 6SE
Telephone 01-969 4727

U.S. Edition published by
Plenum Publishing Corporation
227 West 17th Street
New York 10011

Set in cold type by Academic Industrial Epistemology, London
Printed in Malta by St Paul's Press Ltd

LIST OF CONTRIBUTORS

(Numbers in parentheses indicate the pages on which the authors' contributions begin)

E. Antonini
Università di Roma
Istituto di Chimica
Facoltà di Medicina e Chirurgia
Città Universitaria
00185, Rome, Italy (155)

J. C. Bailar, Jr.,
Department of Chemistry
University of Illinois
Urbana
Illinois 61801
U.S.A. (177)

F. Basolo
Ipatieff Catalytic Laboratory
Northwestern University
Department of Chemistry
Evanston, Illinois 60201
U.S.A.

E. Bayer
Lehrstuhl für Organische Chemie
Universität Tübingen
74 Tübingen
Wilhemstrasse 33, Germany (149)

W. Beck
Institut für Anorganische Chemie
der Universität
8-München 2
Meiserstrasse 1, Germany (149)

S. J. Benkovic
Department of Chemistry
Pennsylvania State University
152 Davey Laboratory
University Park, Pennsylvania 16802
U.S.A. (43, 131)

J. Bjerrum
Department I
Inorganic Chemistry
H. C. Ørsted Institute
Copenhagen, Denmark (131)

G. C. Bond
Department of Industrial Chemistry
Brunel University
Kingston Lane
Uxbridge, Middlesex
England (141)

M. Boudart
Department of Chemical Engineering
Stanford University
Stanford, California 94305
U.S.A. (149)

R. C. Bray
School of Molecular Sciences
University of Sussex
Falmer, Brighton BN1 9QJ
Sussex, England (131)

H. H. Brintzinger
Universität Konstanz
Fachbereich Chemie
775 Konstanz, Postfach 733, Germany. (149)

R. L. Burwell, Jr.,
Ipatieff Catalytic Laboratory
Department of Chemistry
Northwestern University
Evanston, Illinois 60201
U.S.A. (51, 141)

F. Calderazzo
Istituto di Chimica Generale
Università di Pisa
Via Risorgimento 35
56100 Pisa
Italy (165)

S. Carrá
Istituto di Chimica Fisica e
Spettroscopia
Università degli Studi Bologna
Viale Risorgimento 4
Bologna, Italy (177)

J. Chatt
The Chemical Laboratory
University of Sussex
Falmer, Brighton BN1 9QJ
Sussex, England (107, 149)

A. Cimino
Università di Roma
III Cattedra di Chimica Generale
ed Inorganica
Città Universitaria
Rome, Italy (155)

J. E. Coleman
Department of Molecular
Biophysics and Biochemistry
Yale University
333 Cedar Street
New Haven, Connecticut 06510
U.S.A. (3, 141)

P. Cossee
Shell Research Ltd
Thornton Research Centre
P.O. Box No. 1
Chester CH1 3SH,
England (141)

D. A. Dowden
Imperial Chemicals Industries Ltd
Agricultural Division
P.O. Box No. 6
Billingham Tees
TS23 1LD, England (165)

G. L. Eichhorn
National Institute of Health
Gerontology Research Center
Baltimore City Hospital
Baltimore, Maryland 21224
U.S.A. (131)

D. D. Eley
Department of Chemistry
University of Nottingham
University Park
Nottingham NG7 2RD
England (131)

C. Franconi
Università di Venezia
Istituto di Chimica Fisica
Campo Celestia, 2737-B
Venice 30122, Italy (141)

J. J. Fripiat
Laboratoire de Physico-Chemie Minerale
42 de Croylaan
3030 Heverlee-Louvain
Belgium (131)

F. G. Gault
Laboratoire de Catalyse
Institut de Chimie
Université Louis Pasteur
1 Rue Blaise Pascal
67-Strasbourg, France (165)

J. E. Germain
Ecole Supérieure de Chimie
Industrielle de Lyon
43 Boulevard du 11 Novembre 1918
69621 Villeurbanne, France (155)

R. D. Gillard
University Chemical Laboratory
University of Kent at Canterbury
Canterbury, Kent
England (155)

M. Graziani
Facoltà di Chimica Industriale
Università di Venezia
Calle Larga S. Marta, 2137
Venice 30100, Italy (177)

M. L. H. Green
Inorganic Chemistry Laboratory
University of Oxford
South Parks Road
Oxford OX1 3QR
England (165)

W. K. Hall
Gulf Research & Development Company Ltd.,
P. O. Drawer No. 2038
Pittsburgh, Pennsylvania 15230
U.S.A. (101, 131)

J. Halpern
University of Chicago
Department of Chemistry
Chicago, Illinois, 60637
U.S.A. (107, 165)

G. A. Hamilton
Department of Chemistry
Pennsylvania State University
152 Davey Laboratory
University Park
Pennsylvania 16802
U.S.A. (47, 165)

P. Henry
Department of Chemistry
Guelph University
Guelph, Ontario
Canada (131)

J. A. Ibers
Department of Chemistry
Northwestern University
Evanston, Illinois, 60201
U.S.A. (149, 155)

B. R. James
Department of Chemistry
University of British Columbia
Vancouver 8, B.C., Canada (141)

C. R. Jefcoate
Department of Biochemistry
University of Wisconsin
Madison, Wisconsin 53706
U.S.A. (155)

K. Jonas
Max-Planck-Institut für Kohlenforschung
433 Mülheim-Ruhr
Kaiser-Wilhelm-Platz 1
Germany (149)

C. Kemball
Department of Chemistry
University of Edinburgh
West Mains Road
Edinburgh EH9 3JJ
Scotland (85, 165)

H. Knözinger
Physikalisch-Chemisches
Institut der Universität
8-München 2
Sophienstrasse 11
Germany (131)

R. J. Kokes
Department of Chemistry
Johns Hopkins University
Remsen Hall
Baltimore, Maryland 21218
U.S.A. (75, 141)

H. L. Krauss
Freie Universität Berlin
Fachbereich 21-Chemie-Institut für
Anorganische Chemie
1-Berlin 33
Fabeckstrasse 34/36
Germany (149)

R. Lontie
Laboratorium voor Biochemie
Katholieke Universiteit te Leuven
Degenstraat 6
B-3000 Leuven
Belgium (155)

P. M. Maitlis
Department of Chemistry
The University
Sheffield S3 7HF
Yorks, England (131)

J. Manassen
Department of Plastics Research
Weizmann Institute of Science
Rehovoth, Israel (177)

A. Marani
Facoltà di Chimica Industriale
Università di Venezia
Calle Larga S. Marta, 2137
Venice 30100, Italy

J. C. Marchon
Department of Chemistry
Stanford University
Stanford, California, 94305
U.S.A. (149)

A. Martell
Department of Chemistry
Texas A & M University
College Station
Texas 77843
U.S.A. (155)

K. Mosbach
Chemical Centre, Biochemistry 2
Lund Institute of Technology
P.O. Box No. 740
S-220 07, Lund 7
Sweden (177)

G. W. Parshall
E. I. du Pont de Nemours & Company
Central Research Department
Experimental Station
Wilmington, Delaware 19898
U.S.A. (165)

R. C. Pitkethly
B.P. Research Centre
Basic Research Projects
Chertsey Road
Sunbury-on-Thames
Middlesex, England (177)

E. K. Pye
Department of Biochemistry
Medical School
University of Pennsylvania
Philadelphia, Pennsylvania, 19104
U.S.A. (177)

A. Rigo
Università di Venezia
Istituto di Chimica Fisica
Campo Celestia, 2737-B
Venice 30122
Italy (131)

J. J. Rooney
Department of Chemistry
David Keir Building
Stranmillis Road
Queen's University
Belfast BT9 5AG
Northern Ireland (177)

S. J. Teichner
Université Claude Bernard
Lyon I, and
Institut de Recherche
sur la Catalyse,
69 Villeurbanne, France (91, 165)

R. Ugo
Istituto di Chimica Generale
Via Venezian 21
Milano 20133
Italy (155)

V. Ullrich
Physiologisch-Chemisches Institut der
Universitat des Saarlandes
665 Hamburg
Germany (155)

L. Vaska
Department of Chemistry
Clarkson College of Technology
Potsdam, New York 13676
U.S.A. (141)

C. Veeger
Department of Biochemistry
Agricultural University
Wageningen
The Netherlands (1, 149)

J. J. Villafranca
Department of Chemistry
Pennsylvania State University
212 Whitmore Laboratory
University Park
Pennsylvania 16802
U.S.A. (141)

H. C. Volger
Koninklijke/Shell Laboratorium
Badhuisweg 3
Amsterdam
The Netherlands (165)

D. D. Whitehurst
Research and Development
Mobil Oil Corporation
Paulsboro', New Jersey 08066
U.S.A. (177)

PREFACE

This conference on Catalysis was held under the auspices of the NATO Science Committee as part of its continuing effort to promote the useful progress of science through international cooperation.

The Science Committee Conferences are deliberately designed and structured to focus expert attention on what is not known, rather than what is known. The participants are carefully selected to bring together a variety of complementary viewpoints. Through intensive group discussion, they seek to reach agreement on conclusions and recommendations for future research which will be of value to the scientific community.

We believe that the endeavour has been particularly successful in the present case. Some twenty-five papers, either in the form of reprints or specially written reviews were contributed by the participants for advance circulation, to outline the state-of-the art in the three areas of heterogeneous, homogeneous and metalloenzyme catalysis and to focus attention on key problems. The availability of this background material precluded the need for lengthy introductory presentations and permitted rapid initiation of interdisciplinary discussions. All participants gave generously and enthusiastically of their expertise and effort during the week of the meeting, often long past normal bedtime hours, and we extend to them our deep gratitude.

We are delighted to have this opportunity to record our special thanks to Professor Fred Basolo and Robert Burwell for their diligent efforts as Chairman and Co-Chairman of the meeting, to their colleagues on the Organizing Committee, Professors S. Benkovic, J. Chatt, C. Kemball, R. Ugo and V. Ullrich, for their wise guidance, and to the leaders and recorders of the Working Groups, as listed, for their indispensable dedication. The assistance of Professor R. Lontie in preliminary editing of this volume is gratefully acknowledged.

We also wish to record our appreciation to the US Office of Naval Research, which contributed significantly to the success of the meeting through provision of logistic support.

Eugene G. Kovach
Deputy Assistant Secretary General for Scientific Affairs

INTRODUCTION

The desirability of classifying problems and accelerating their solution in the general area defined as catalysis has become apparent to many within the scientific community. The field concerned with catalysis can be generally divided into three specialities which encompass enzymatic, heterogeneous, and homogeneous reactions.

Heterogeneous catalysis is still the basis of most commercial chemical processes and therefore is the subject of extensive investigation within industry. Although homogeneous catalysis has not gained a similar state of commercial importance, basic research on many of the mechanisms is in progress. The third area, enzymatic catalysis, has developed rather independently mainly through the efforts of biochemists and biophysicists.

Although rapid progress in all three specialities is evident, it is also obvious that significant developments within one have not always been recognized for their pertinence to another. It was the intent of the Organizing Committee, therefore, to assemble a group of active scientists knowledgeable in contemporary theory and experimental design within their respective specialities to foster a mutually beneficial exchange of information. Such communication might particularly lead to recognition of problems that are truly interdisciplinary in scope and also to proposed solutions for these problems.

To enhance the degree of interaction six topics, "Kinetics and Mechanism" "Hydrogenation/Dehydrogenation", "Nitrogen Fixation", "Oxygen Activation" "Hydrocarbon Activation" and "Heterogenizing Catalysts", were chosen for discussion, based mainly on the judgement of the Organizing Committee that the probability for overlap between the above three specialities was maximal for these subjects. It is emphasized that these topics served merely as a framework for discussion and were neither all inclusive nor representative of other equally important problems within the field of catalysis. The resulting reports were produced during three days of intensive discussion within small groups that followed a short period of general introduction to acquaint all with the semantics and present status of research within the three subdivisions of catalysis.

By the nature of the structure of the conference, no *systematic* consideration was given to the problems which lie within just one of the areas of catalysis nor to overlap between any of these areas and external areas—for example, between enzyme catalysis and microbiology, or between heterogeneous catalysis and surface chemical physics. Furthermore, the overlap between metalloenzymes and organometallic chemistry was considered extensively but many aspects of coordination chemistry potentially useful for catalysis of such reactions as hydrolysis, transamination, decarboxylation, etc., received little attention. Thus, the conclusions of the conference do not extend to aspects primarily within each speciality nor to overlaps between any one area and adjacent non-catalytic areas.

It was impossible within the allotted time period to correct imbalance as to the length of individual reports of the six working groups. Such differences should not be construed as reflecting the relative importance or priority to be assigned a given topic, but simply as a consequence of the inequality in topic scope and the heterogeneity of the groups involved. It seemed preferable to reproduce the group reports without attempting to edit them or reduce them to a uniform format as this might have destroyed their originality. Instances of repitition function to pinpoint problems, techniques, and recommendations which are truly interdisciplinary. The references at the end of the Group Reports are intended to give the reader a general guide to further information on the subject.

The present volume includes, in addition to the Group Reports, specially commissioned papers reflecting the current situation on key aspects of catalysis research. This mixing of scientists from related but not identical disciplines led to an enthusiastic articulation of potential problem areas, present deficiencies and self-criticism within and without their own areas of speciality. The Conclusions (Page 187) attempt to summarize the findings of the Working Groups, but a close reading of the Group Reports as well is strongly urged.

The Organizing Committee

ACKNOWLEDGEMENTS

Plenum Publishing Company would like to thank the respective authors and publishers for permission to reproduce material from the following sources:

THE AMERICAN CHEMICAL SOCIETY

Breslow, R. and Overman, L. E., *J. Am. Chem. Soc.*, (1970), **92**, 1075, Figure 1.

Burwell, R. L. Jr., The Mechanism of Heterogeneous Catalysis, *Chemical and Engineering News*, August 22, 1966.

US NATIONAL ACADEMY OF SCIENCES

Reeke, G. N., Hartsuck, J. A., Ludwig, M. L., Quiocho, F. A., Steitz, T. A. and Lipscomb, W. N., *Proc. Nat. Acad. Sci. US*, (1967), **58**, 2220.

Bachmayer, H., Piette, L. H., Yasunobu, K. T. and Whiteley, H. R., *Proc. Nat. Acad. Sci. US*, (1967), **57**, 122.

Herskovitz, T., Averill, B. A., Holm, R. H., Ibers, J. A., Phillips, W. D. and Weiher, J. F., *Proc. Nat. Acad. Sci. US*, (1972), **69**, 2437.

ANNUAL REVIEWS INC.

Blow, D. M. and Steitz, T. A., *Annual Reviews of Biochemistry, Volume 30*, 1970, Figure 5.

ACADEMIC PRESS, NEW YORK.

Hartsuck, J. A. and Lipscomb, W. N., *The Enzymes, Volume 3* (3rd edition), 1971, Figure 14.

Morooka, Y., and Osaki, A., *J. Catalysis*, (1966), **5**, 116, Figure 4.

Winter, E. R. S., *J. Catalysis*, (1969), **15**, 144, Figure 6.

Winter, E. R. S., *J. Catalysis*, (1970), **19**, 32, Figure 5.

Winter, E. R. S., *J. Catalysis*, (1971), **22**, 158, Figure 6.

McAffrey, E. F., Klissurski, D. G. and Ross, R. A., *J. Catalysis*, (1972), **26**, 380, Figure 4.

ELSEVIER PUBLISHING COMPANY

Lindskog, S. and Nyman, P. O., *Biochim. Biophys. Acta*, (1964), **85**, 462, Figure 5.

MACMILLAN JOURNALS LIMITED

Liljas, A., Kannan, K. K., Bergsten, P-C., Waara, I., Fridborg, K., Strandberg, B., Carlbom, U., Järup, L., Lövgren, S. and Petef, M., *Nature New Biology*, (1972), **235**, 131, Figures 3, 6 and 7.

PERGAMON PRESS, NEW YORK

Palmer, G., Brintzinger, H., Estabrook, R. W. and Sands, R. H., *Magnetic Resonance in Biological Systems* 1967 page 159.

JOHN WILEY AND SONS, NEW YORK

Coleman, J. E., *Progress in Bioorganic Chemistry Volume 3*, 1971, page 250.

COLD SPRING HARBOR LABORATORY

Watenpaugh, K. D., Sieker, L. C., Herriott, J. R. and Jensen, L. H., *Cold Spring Harbor Symposium on Quantitative Biology*, (1971), **36**, 359.

Carter, C. W., Freer, S. T., Xuong, N. H., Alden, R. A. and Kraut, J., *Cold Spring Harbor Symposium on Quantitative Biology*, (1971), **36**, 381.

THE BIOCHEMICAL SOCIETY

Jensen, L. H., *Abstracts of Communications, Biochemical Society, Volume 1*, 1973, page 1.

SOCIETE DE CHIMILE, FRANCE

Germain, J. E. and Laugier, R., *Bull. Soc. Chim. France*, (1972), page 54.

SOCIETE DE CHIMIE PHYSIQUE, FRANCE

Vijh, A. R., *J. Chim. Phys.*, (1972), **69**, 1695.

CONTENTS

Contributors v
Preface ix
Introduction xi
Acknowledgements xiii
Contents xv

Part 1 Metalloenzyme Catalysis

General Survey
C. Veeger 1

Catalysis by Metalloenzymes
J. E. Coleman 3

Metal Ion Activated Enzymes
S. J. Benkovic 43

Redox Reactions Catalysed by Metalloenzymes
G. A. Hamilton 47

Part 2 Heterogeneous Catalysis

General Survey
R. L. Burwell Jr. 51

The Nature of Active Sites
R. J. Kokes 75

Selectivity and Poisoning
C. Kemball 85

Oxidation and Deoxidation
S. J. Teichner 91

Reaction Schemes and Coordinates
W. K. Hall 103

Part 3 Homogeneous Catalysis

Homogeneous Catalysis by Metal Ions and Complexes
J. Chatt and J. Halpern. 107

Part 4 Working Group Reports

Kinetics and Mechanisms
W. K. Hall et. al. 131

Hydrogenation/Dehydrogenation
G. C. Bond et al. 141

Nitrogen Fixation
J. Chatt et al 149

Oxygen Activation
C. R. Jefcoate et al 155

Hydrocarbon Activation
G. W. Parshall et al 165

Heterogenizing Catalysts
J. Manassen et al 177

Conclusions 187

Subject Index 189

PART I
METALLOENZYME CATALYSIS

GENERAL SURVEY OF ENZYMOLOGY

C. Veeger
Department of Biochemistry, Agricultural University, Wageningen, The Netherlands

Enzyme research started in the 1930s via two lines of development. Firstly, via the experience that enzymes as proteins can be crystallized and obtained in more or less pure form and are solely responsible for the catalytic activity of biological reactions. Secondly, via the observation that the activity, as well as being rectilinearly dependent on the enzyme concentration, followed a Langmuir saturation isotherm in its dependence on the substrate concentration. The 1940s and part of the 1950s were mainly devoted to the discovery of new enzymes which all possessed in general the above-mentioned characteristics, but new concepts were also introduced. Expressions like V_{max} for maximum rate at substrate saturation, and K_m for the Michaelis constant (identical with the concentration giving activity 50% of maximal activity) were introduced. In addition it was found that enzymes can be inhibited by compounds resembling the substrate in structure. Other expressions like competitive inhibition, i.e. inhibition by competing with the substrate for the active site, and non-competitive inhibition, i.e. by exerting influence on the maximal rate (rate-determining step), were introduced. But still most of the results were scarcely concerned with the chemical nature of the concept enzyme-substrate complex.

The real breakthrough came along several lines:

(1) Distinguishing the different kinetic mechanisms in enzymatic reactions.

(2) The three-point binding attachment of substrate to the active site of the enzyme in order to account for the observed stereospecificity of enzymatic reactions.

(3) The translation of the concept of enzyme-substrate complex for peroxidase in terms of valency changes of heme Fe.

(4) The concept of induced fit, postulating that the active centre of enzyme catalysis adjusts itself to the molecular dimensions of the substrate upon binding.

(5) The discovery that redox reactions mainly proceed not by radical mechanisms as formerly postulated, but by H^--transfer.

(6) The application of flow techniques to the study of enzyme-substrate binding.

The results point to the fact that enzymes form upon reaction with their substrate an intermediate (by physisorption) prior to reaction (by chemisorption).

In the last decade enzymology was strongly stimulated by the elucidation of the three-dimensional structure of enzymes by X-ray diffraction. Its application to enzyme-substrate analogue complexes provided for the first time insight into where the active site was located and how catalysis could be promoted by vicinal groups at the active center. In addition, new concepts were introduced in attempts to explain the regulation of enzyme activity in the cell. Every biochemist is nowadays familiar with expressions such as allosteric site — a binding site, different from the active site where a regulator compound exerts its activating or inhibiting effect on the enzyme activity. In addition it was observed that not all enzymes catalyse their reaction according to a Langmuir-isotherm saturation curve (so-called allosteric catalysis). Furthermore, evidence was obtained that conformational changes of the protein occur during catalysis and allosteric control.

The progress in enzymology is to a great extent due to the application of new molecular-physical techniques, like ESR, NMR, Mössbauer spectroscopy, circular dichroism, etc.. The ligand field of unknown Fe complexes in enzymes was elucidated in this way. It must be remarked that although metals play an important role in enzyme catalysis, a large number of enzymes are able to catalyse their reaction in the absence of any metal.

Although it can be stated that our knowledge of enzyme catalysis has definitely increased, it is still not known what factors are essentially responsible for the large increase in rate ($10^{10} - 10^{18}$) over the non-catalysed reaction. It is considered at present that, rather than a reduction in activation energy, the increase in entropy caused by distortion of the substrate upon binding is a main factor. Furthermore, it is unknown what influence the polarity of the micro-environment at the active site of the enzyme has on the rate of the reaction. Theories have been available for many years. In the 1900s the lock-and-key hypothesis of enzyme action was introduced and accepted until it was visualized that conformational changes during catalysis occur. In other words, theories are readily available and can even be favoured for many years; but the final word is given by the experiment. Without doubt our knowledge of enzymatic catalysis will contribute to developments in and understanding of heterogeneous and homogeneous catalysis.

CATALYSIS BY METALLOENZYMES

Joseph E. Coleman

Department of Molecular Biophysics and Biochemistry, Yale University, New Haven, Connecticut 06510, *USA*

Catalysis by metalloenzymes occupies a relatively unique position when placed in a general consideration of heterogeneous and homogeneous catalysis. The distinguishing feature of course is that even if one tends to focus on the metal ion as essential to the catalytic center, one must remember that this center carries with it a large protein molecule of molecular weight from 6000 in the simplest cases to over a 100 000 in the more complex metalloenzymes. The primary, secondary, and tertiary structure of the macromolecule surrounding the metal ion, resulting from the folding of the amino-acid polymer (and believed to be dictated by the primary sequence of the 20 different amino acids), makes possible an enormous variation in the microenvironment of the metal ion. This microenvironment can also determine the reactivity of the organic groups (the various amino-acid side-chains) that can assume a role in a given catalytic reaction in addition to the metal ion. Such variation is not readily achieved in small molecular systems. Since the great specificity and enormous rate enhancements at 37 °C characteristic of biological catalysis are a property of catalytically active protein molecules, many of which do not contain metal ions, the chemistry of metalloenzymes must be considered as a special case of enzymatic catalysis.

While the metal may enable the enzyme to carry out a particular specialized reaction, the mechanism will undoubtedly involve the metal and a particular protein microenvironment or reactive group(s) as joint participants in the catalytic event. Metal participation may be an absolute requirement for some reactions, in the sense that nature has not evolved another mechanism. Certain oxidation-reduction reactions might fall in this category. On the other hand for some important biological reactions there are both metalloenzymes and non-metalloenzymes catalysing similar reactions, hence nature has been able to evolve a purely organic system to achieve the desired catalytic mechanism. Therefore it may not be too surprising if catalysis by metalloenzymes turns out to have as many if not more features characteristic of protein catalysis in general as it does of metal catalysis in the simpler chemical systems where the metal carries most of the catalytic function albeit modified in some instances by the electronic properties of small ligand molecules.

Metalloenzymes or metal-activated enzymes catalyse an enormous variety of organic reactions. If one scans the Table of enzymes compiled by the Enzyme Commission[1] and grouped according to the type of reaction catalysed, metalloenzymes are not restricted to any particular group, but appear as catalysts for all types of reactions. Thus neither the presence of the metal nor the reaction-type seems to be restrictive as far as metal-assisted enzyme catalysis is concerned with the possible exception of oxidoreductases not involving coenzymes. These appear to be exclusively metalloenzymes where the metal appears in most cases to function as an electron acceptor or donor. The variety of reactions catalysed by metal-containing enzymes is indicated in Table I by a few selected examples.

TABLE I

Examples of Reactions Catalysed by Metalloenzymes

Enzyme	Mol. wt.	Metal per molecule	Cofactor per molecule	Reaction
Carbonic anhydrase	30 000	1 Zn	None	$CO_2 + H_2O \rightleftharpoons H^+ + HCO_3^-$
Carboxy-peptidase A	34 500	1 Zn	None	Hydrolyses C-terminal peptide bonds of proteins and peptides when the side-chain is aromatic or branched aliphatic
Alkaline phosphatase	80 000	2 Zn	None	$ROPO_3^{2-} + H_2O \rightleftharpoons ROH + HPO_4^{2-}$
Alchohol dehydrogenase	84 000	2-4 Zn	NAD	$RCH_2OH + NAD^+ \rightleftharpoons RCHO + NADH + H^+$
Tyrosinase	120 000	4 Cu	None	Oxidizes phenols to the *o*-dihydric compound, oxidizes catechols to the *o*-quinones, O_2 as oxidant
Nitrogenase	Complex of two proteins			
	55 000	4 Fe	4 labile S^{2-}	Catalyses reduction of N_2 (acetylene, cyanide and protons)
	220 000	2 Mo 20 Fe	20 labile S^{2-}	
Superoxide dismutase	34 000	2 Cu 2 Zn	None	$O_2^- + O_2^- + 2H^+ \rightleftharpoons H_2O_2 + O_2$
Xanthine oxidase	300 000	8 Fe 2 Mo	2 FAD	Xanthine + $O_2 \rightleftharpoons$ urate + H_2O_2
Ferredoxin	6 000	8 Fe	8 labile S^{2-}	Electron acceptor with photo-activated chlorophyll as donor, reductant in formation of acetyl-CoA from pyruvate in anaerobic bacteria
Cytochrome *c*	12 400	1 Fe	1 heme	Electron acceptor and donor in mitochondria, substrate for cytochrome oxidase
Methylmalonyl-CoA isomerase		Co	B_{12} coenzyme	$CH_3CH(CO_2^-)(CO\text{-}SCoA) \rightleftharpoons {}^-O_2C\text{-}CH_2\text{-}CH_2\text{-}CO\text{-}SCoA$

We still do not have enough information about any single enzyme to understand its mechanism of action in detail. Some general statements about enzyme catalysis are possible and apply equally well to metalloenzymes. The striking features of enzyme catalysis are its great specificity and its great rate enhancement. The specificity is generally considered to be associated with a binding site on the protein whose molecular architecture is tailored to result in special affinity for the substrate (often only one stereoisomer of the substrate) undergoing chemical change; hence the formation of the classical enzyme-substrate complex or Michaelis complex. Binding constants characterizing such interaction, however, span many orders of magnitude. This specificity is not always associated with extraordinary affinity. Such variation may reflect relative substrate concentrations in the biological environment. The actual catalytic step is then visualized as proceeding through the operation of specific and reactive 'catalytic' groups at the substrate-binding or 'active site' of the enzyme. However, as Jencks[2] has pointed out, it may not be possible to separate artificially the two features, since binding is an inseparable part of the catalytic process and it may in the end not be possible to separate certain protein-substrate interactions as primarily 'binding' from those that are primarily 'catalytic'. It is often stated that the enzyme-substrate complex must have some of the electronic characteristics required of the transition-state for the given reaction. The binding energy would then be closely associated with the catalytic step.

As Jencks[2] has expressed it, aside from theorizing, there are two ways of approaching the mechanism of enzyme action; by examining the properties of the enzymes or by examining the nature of chemical reactions and their catalysis. There has been lively discussion as to the relative value of these two approaches. It is very difficult to get the kind of detailed chemical information on an enzyme that is needed to answer questions about the mechanism of reaction. On the other hand catalysis in simple non-enzymatic systems is often difficult to compare to the analogous enzyme catalysed reaction. The latter often proceeds by a different pathway than that taken by the model reaction because the enzyme makes possible a mechanism in which the free energy of activation in the absence of enzyme is simply too high to make it the favored path in the non-enzymatic case. Rate laws for the two reactions may also be different in that the enzyme reaction proceeds through the first-order breakdown of the enzyme-substrate complex at neutral pH, while the model may require acid or base catalysis and follow a higher-order rate law. For these reasons it is hard to obtain estimates of the actual magnitude of the rate enhancement which occurs in the case of a given reaction when it is enzyme catalysed as opposed to catalysis of the same reaction in a non-enzymatic system using acids, bases, metals, or small molecules as catalysts. Estimates of this enhancement factor, however, are as high as 10^{14}. In some cases it may be even higher, since the enzyme mechanism may use a pathway that, uncatalysed, is even less favored than the one observed in the model system.

It is generally accepted that formation of the enzyme-substrate complex reduces the free energy of activation of the ensuing reaction. The degree of this reduction, what part is made up by a decrease in enthalpy and what part by a change in entropy, is difficult to answer. Mechanisms by which the fall in free energy is accomplished are usually stated in terms of approximations of the reactants, covalent catalysis, general acid-base catalysis, induction of distortion or strain in the substrate, the enzyme, or both, or a combination of these. Changes in solvent structure near the active center of the enzyme may also be involved.

In this connection it should be noted that one of the advantages of the large protein molecule as catalyst may be that a particularly unstable configuration of amino acids (considered as an isolated structure) may be maintained at one site in the molecule (e.g. the active site) by the stabilizing effect of structure in the rest of the molecule. Thus while the total protein structure would be maintained at a potential minimum (or at least a potential that would ensure relative stability) a highly unfavorable configuration might be maintained in a small area essential to catalysis. This may be pertinent to metalloenzyme catalysis where highly distorted metal-ligand geometries (both bond lengths and bond angles) may be maintained which would be considered highly unstable were the ligands free to adjust themselves in the most stable configuration relative to the metal ion. Such 'strain' could radically affect the reactivity of the metal ion toward a substrate in a mixed complex formed during catalysis.

There are a number of formal analogies between enzymatic and non-enzymatic catalysis. It is not yet known, however, how many of these will have detailed bearing on the mechanism of enzyme catalysis. Heterogeneous catalysis is associated with the attachment of the substrate to the catalyst and is subject to variables such as the lattice spacing in the catalyst as well as electronic factors in substrate and catalyst. A certain amount of stereospecificity can also be invoked since some reactions such as polymerization can be stereospecific.

In homogeneous catalysis there are many more specifically identifiable reactions and catalytic species, and it is among the metal catalysed reactions in this group that models for enzymatic catalysis have been sought. This group includes the very interesting examples of catalysis involving the formation of metal-carbon bonds which include such reactions as olefin oxidation, insertion of CO groups or rearrangement of C-C bonds. In general this class of reactions grouped under the heading of organometallic chemistry does not as yet appear to have many analogies in biological systems with the exception of enzymes using the cobalamins as cofactors[3,4]. In this case the Co in vitamin B_{12} does form C-Co bonds which are important to the mechanism of action.

There is a great variety of reactions catalysed by metal ions which fall in the category of acid-base catalysis. While these include hydrolysis, hydration, condensation, and substitution reactions, only two examples will be given here to illustrate metal-catalysed hydrolysis reactions. A number of ester hydrolyses including the hydrolysis of amino-acid esters have been observed to be catalysed by aqueous solutions of metal salts. The reactions have all been postulated to involve a polarisation of the carbonyl group by formation of a complex between the metal ion and the oxygen atom.

$$-\overset{\overset{\displaystyle O}{\|}}{C}-OR + M^{2+} \underset{}{\overset{+H_2O}{\rightleftharpoons}} -\overset{\overset{\displaystyle M^{2+}}{|}}{\overset{\displaystyle O^-}{\underset{\underset{\displaystyle H\quad H}{\displaystyle O}}{\overset{|}{\underset{\uparrow}{C}}}}}-OR \overset{-H^+}{\rightleftharpoons} -\overset{\overset{\displaystyle M^{2+}}{|}}{\overset{\displaystyle O^-}{\underset{\underset{\displaystyle OH}{|}}{\overset{|}{C}}}}-OR \rightleftharpoons -\overset{\overset{\displaystyle O}{\|}}{C}\diagdown_{OH} + ROH + M^{2+} \quad (1)$$

The metal is polarizing the carbonyl first because of its ability to form a complex and second through its action as a Lewis acid in withdrawing electrons from the carbonyl oxygen. While this model reaction may have a number of structural as well as functional analogies to some enzymatic hydrolysis and hydration reactions (see carboxypeptidase and carbonic anhydrase, below), the rate enhancements caused by the aquo cations are orders of magnitude less than the rate enhancements observed in catalysis of similar reactions by enzymes. Other factors must be involved in the enzymatic case. Possibilities which suggest themselves are: (1) The metal may participate as a Lewis acid along with another protein group which might facilitate general base catalysis or act as a specific nucleophile. The initial polarization by the metal could greatly assist the latter process. (2) The particular orientation of the bound substrate relative to the metal ion may force a type of complex that is either not stable in solution, or is only one among several occurring in solution. This orientation may force a particular stereochemistry on the complex that would not be assumed if the ligand were free to move relative to the metal ion. (3) The binding step may induce a strain in the substrate which has no counterpart in the simple system.

Such considerations suggest that more suitable models of metalloenzyme catalysis might be obtained if additional functional groups besides a metal coordination center were built into the model. In the right circumstances certain stereochemical restraints might also enhance catalysis. There is some evidence that such features do result in enhanced catalysis in model systems and a couple of examples will illustrate this point. Copper(II) brings about a dramatic rate acceleration of phosphate monoester hydrolysis when the substrate is 2-(4(5)-imidazolyl) phenyl phosphate[5]. Two possible complexes formed between Cu(II) and this ligand are illustrated in I and II. The enhancement for hydrolysis of the P-O bond in the complex is

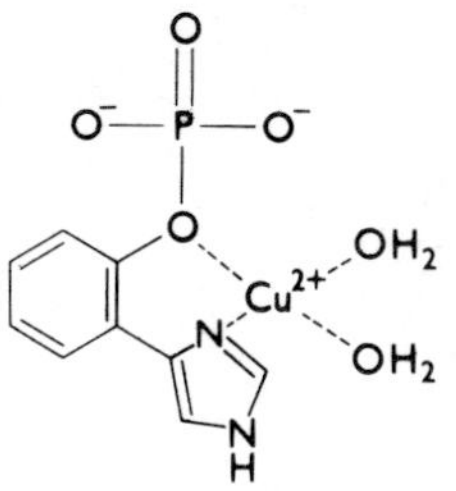

I

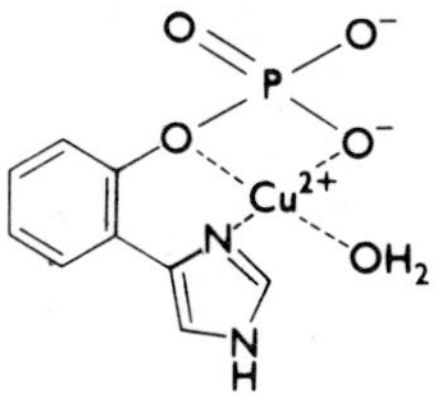

II

approximately 10^4 at pH 6 compared to the uncatalysed rate. This is at least 1-2 orders of magnitude larger than rate enhancements previously observed for aquo cations catalysing simple esterolytic systems, particularly those involving phosphate esters. Picturing either complex I or II as the transition state for the reaction, there are several possible explanations for this particularly efficient catalysis. The presence of the imidazole nitrogen tends to force complex formation with the ether oxygen rather than the phosphoryl oxygen; the latter is preferred in an ester containing only the phosphate dianion as the coordinating center. In I, the metal ion acts as a more effective acid catalyst than a proton, lowering the pK_a of the leaving group so that facile hydrolysis of the dianion may be observed (generally encountered only with leaving groups of $pK_a < 7$ if the usual metaphosphate route is assumed). In intermediate II, the Cu(II) may also serve to induce strain in the P-O bond and/or partially neutralize charge on the phosphate, leading to a nucleophilic displacement by solvent on phosphorus. Hydrolysis of II via nucleophilic attack by water is kinetically indistinguishable from hydroxide attack on the monoanion of I.

While this system brings up the question of nucleophilic attack in concert with metal-promoted polarization, solvolysis of this system in mixed methanol-water using both Cu(II) and the uncatalysed system show the same methylphosphate inorganic phosphate ratio. Furthermore the Cu(II)-assisted hydrolysis in $H_2{}^{18}O$ shows no detectable incorporation of ^{18}O into the phosphate[5]. Both findings are against the nucleophilic mechanism and would suggest that the metal via the directed coordination at the ester oxygen potentiates the usual metaphosphate mechanism primarily through its function of electron withdrawal, but perhaps also by inducing strain at the P-O ester bond. Some of these considerations will be pertinent to the mechanism of catalysis by alkaline phosphatase (see below).

This very brief discussion of models will be concluded with the mention of one additional attempt to mimic enzyme catalysis with a model system. While some model metal-catalysed systems in which the substrate binds strongly to the metal ion have achieved catalytic enhancements of 10^9, it will be clear below that the type of interaction between substrate and metal in many metalloenzymes appears to be of a type, often monodentate, that would not suggest a particularly stable complex in solution. Thus as has been noted before, metalloenzymes would appear to bind their substrates principally by use of the same kinds of forces, including hydrophobic interactions, common to all enzymes[6]. These considerations led Breslow and his colleagues[6] to design a model in which they attached a metal-binding group as a catalytic center to a large molecule, a cyclodextran, with a hydrophobic cavity as a substrate-binding site (Fig. 1).

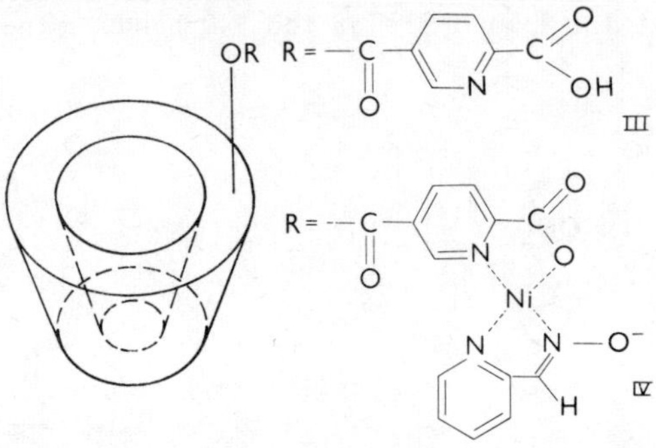

Fig. 1. Catalyst for ester hydrolysis produced by attaching a Ni^{2+} chelate to cyclodextran ring.

The parent compound III was produced by acylation of the secondary hydroxyl group of the cycloamylose. Addition of 1 equiv of $NiCl_2$ to III produced a complex with an absorption spectrum characteristic of Ni picolinates. This was converted to IV by the addition of 1 equiv of pyridine carboxaldoxine (PCA). The catalysis of the hydrolysis of *p*-nitrophenyl acetate (PNPA) by IV was then determined. The *p*-nitrophenyl group has been shown to bind in the hydrophobic cavity. Hydrolysis of PNPA catalysed by PCA-Ni^{2+} has shown a 2-step process-acetylation of the PCA oxygen, followed by metal-catalysed hydrolysis of PCA acetate. It is the first step of this reaction which should be facilitated by IV. The k_{obsd} for the uncatalysed reaction at pH 5 $^{25°C}$ is 7.1×10^{-5} min^{-1}; for III Ni^{2+}, 2.2×10^{-4} min^{-1}; for PCA-Ni at 10 mM, 5×10^{-2} min^{-1}; and for IV it is 19×10^{-2} min^{-1}. Thus the addition of the hydrophobic binding site increases the catalytic rate four times. Competition experiments show inhibition by agents competing for binding in the cavity. The increase in the initially relatively effective catalysis of the functional group, PCA-Ni^{2+}, has been increased only modestly by the addition of the hydrophobic cavity. Models of IV show, however, that while the PCA oxygen can reach the acetyl group of bound PNPA, several degrees of rotational freedom must be frozen for this to occur. As Breslow points out, if greater rigidity were built into the catalyst this problem should decrease. While rotational degrees of freedom are undoubtedly extremely limited in metalloenzyme-substrate complexes, some of this rigidity may be obtained only after conformational changes in certain functional groups of the protein which were relatively free to rotate prior to substrate binding (see below).

The precise features of catalysis by metalloenzymes will probably differ in some respects from even the best small models we can devise. However, as we learn more about the macromolecules it may be possible to introduce features in small molecular complexes that may better mimic specific properties, either specificity or rate enhancement, demonstrated by the metalloprotein catalysts.

With the enormous variation in protein structure and function which we know to be possible, and in view of the broad spectrum of organic reactions catalysed by metalloenzymes, it is difficult as yet to discover many general principles when discussing the specific aspects of metalloenzyme catalysis. The subject is probably better approached in the context of the present broad-ranging consideration of catalysis by taking up a limited number of examples of metalloenzymes for which we can discuss mechanism on the basis of rather detailed structural and functional information about the molecule.

Knowledge of the structure of proteins has taken a sudden jump with the advent of X-ray crystallography of proteins. It should be stated at the outset that in terms of the information needed to discuss the mechanism of enzyme action in solution, an X-ray structure of a protein does not provide an unequivocal solution. Determination of the structure of an active enzyme-substrate complex is not possible, and deductions must be made from complexes with very poor substrates or nonactive substrate analogues. Resolution is seldom as good as in the case of crystallography of small molecules; hence, the fine variations in bond angles or distances that will be so important to the actual chemical reaction are not discernible. The static average structure discernible by X-ray may or may not deviate from the dynamic state in solution. Finally the question of conformational differences between solution and the solid state are matters of active investigation. Perhaps, since protein crystals are 50% water, the crystalline structure of the relatively rigid protein may not vary too much from the one in solution, but whether it is the only one possible in solution is an unanswered question. It is worth emphasizing, however, that the availability of a high-resolution X-ray structure for a given enzyme coupled with detailed data on the solution chemistry advances our knowledge of the system by orders of magnitude. The focus on possible mechanisms becomes much sharper. Thus wherever possible I will confine my discussion to systems for which both types of information are available.

There are now a number or metalloenzymes for which high resolution X-ray structures are either complete or in progress (Table II). For the reasons outlined above I will confine most of my detailed discussion to enzymes from this group. While this group contains what I think are representative examples, it is in a sense artificially limited in the sense that many complex metalloenzymes or arrays of metalloenzymes as in the mitochondrial respiratory chain are not as yet amenable to structural determination on the level possible with those enzymes listed in Table I. The former represent many highly important biological functions. At our present state of knowledge, however, it is hoped that confining the discussion to the well-characterized systems will gain in detailed chemistry what it lacks in biological completeness. The particular systems I will discuss are: (1) carboxypeptidase A as an example of metallo-peptidases; (2) carbonic anhydrase as an example of metal-assisted hydration where the simplicity of the reaction, hydration of CO_2, brings up some interesting problems; (3) alkaline phosphatase as an example of phosphate-ester hydrolysis and phosphoryl transfer as well as an illustration of the additional complexity posed by multisubunit enzymes; and (4) ferredoxins as an example of metalloenzymes participating in oxidation-reduction. A brief comparison of the latter to a heme enzyme, cytochrome *c*, will be made.

CARBOXYPEPTIDASE A

Carboxypeptidase A is among the peptidases found in the exocrine secretion of the mammalian pancreas. The bovine enzyme is a zinc metalloenzyme containing 1 mol of Zn per mol of enzyme, mol. wt. 34 500[7]. Zinc is essential

TABLE II

Metalloenzymes for Which Crystal Structures Have Been Obtained

Enzyme	Metal	Resolution in Å	Reference
Carboxypeptidase A	Zn	2.0	(a)
Carbonic anhydrase C	Zn	2.0	(b)
Thermolysin	Zn	2.3	(c)
Alkaline phosphatase	Zn	7.7	(d)
Alcohol dehydrogenase species	Zn	6.0	(e)
Rubredoxin	Fe	1.5	(f)
Ferredoxin (*Micrococcus aerogenes*)	Fe	2.8	(g)
High-potential iron protein	Fe	2.25	(h)
Aspartate transcarbamylase	Zn	5.5	(i)
Cytochrome *c*	heme Fe	2.45	(j)
Cytochrome b_5	heme Fe	2.0	(k)

(a) HARTSUCK, J.A. and LIPSCOMB, W.N. (1971). *The Enzymes, Vol.3*, (3rd edn.), p.1
(b) LILJAS, A., KANNAN, K.K., BERGSTÉN, P.-C., WAARA, I., FRIDBORG, K., STRANDBERG, B., CARLBOM, U., JÄRUP, L., LÖVGREN, S. and PETEF, M. (1972). *Nature New Biol.*, 235, 131.
(c) MATTHEWS, B.W., JANSONIUS, J.N., COLMAN, P.M., SCHOENBORN, B.P. and DUPOURQUE, D. (1972). *Nature New Biol.*, 238, 37.
(d) HANSON, A.W., APPLEBURY, M.L., COLEMAN, J.E. and WYCKOFF, H.W. (1970). *J. Biol. Chem.*, 245, 4975; KNOX, J.R. and WYCKOFF, H.W. (1973). *J. Mol. Biol.*, 74, 533.
(e) BRÄNDÉN, C.I., ZEPPEZAUER, E., BOIWE, T., SODERLUND, G., SODERBERG, B.-O. and NORDSTROM (1970). *Pyridine Nucleotide-Dependent Dehyrogenases*, (ed. Sund, H.), (Springer-Verlag, New York), p.129.
(f) WATENPAUGH, K.D., SIEKER, L.C., HERRIOTT, J.R. and JENSEN, L.H. (1971). *Cold Spring Harbor Symposia*, 36, 359.
(g) JENSEN, L.H. (1973). *Biochem. Soc. Trans.*, 1, 1.
(h) CARTER, C.W., Jr., FREER, S.T., XUONG, Ng.H., ALDEN, R.A. and KRAUT, J. (1971). *Cold Spring Harbor Symposia*, 36, 381.
(i) WILEY, D.C., EVANS, D.R., WARREN, S.G., McMURRAY, C.H., EDWARDS, B.F.P., FRANKS, W.A. and LIPSCOMB, W.N. (1971). *Cold Spring Harbor Symposia*, 36, 285.
(j) DICKERSON, R.E., TAKANO, T., EISENBERG, D., KALLAI, O.B., SAMSON, L., COOPER, A. and MARGOLIASH, E. (1971). *J. Biol. Chem.*, 246, 1511; TAKANO, T., SWANSON, R., KALLAI, O.B. and DICKERSON, R.E. (1971). *Cold Spring Harbor Symposia*, 36, 397.
(k) MATHEWS, F.S., ARGOS, P. and LEVINE, M. (1971). *Cold Spring Harbor Symposia*, 36, 387.

for activity. Zinc can be reversibly removed with chelating agents. The apoenzyme is inactive. Readdition of 1 mol of Zn per mol of enzyme totally restores activity. The enzyme is synthesized as a part of a zymogen of mol. wt. 87 000[7]. The zymogen, although it contains the Zn atom, is inactive against polypeptide and protein substrates. The large protein apparently contains the carboxypeptidase A molecule as the C-terminal fragment of the zymogen which is released by the extensive hydrolysis of internal peptide bonds accompanying the activation catalysed by trypsin[7,8]. The exact structural changes which 'activate' the active site during this process are presently under investigation[9]. There are some indications that such changes may be relatively slight, since the substrate binding site appears to exist in the proenzyme[9] and some halogenated amino acids are actually hydrolyzed efficiently by the proenzyme[9]. I will not discuss this interesting aspect of the enzymology further, but briefly outline our detailed knowledge of the active enzyme based on the X-ray structure[8], the extensive solution chemistry [7,10] and the primary structure[11].

The reaction catalysed by the enzyme is illustrated by the two classical substrates, benzoyl-glycyl-L-phenylalanine (V) and hippuryl-β-phenyllactate (VI), the analogous synthetic ester. The enzyme is an exopeptidase and

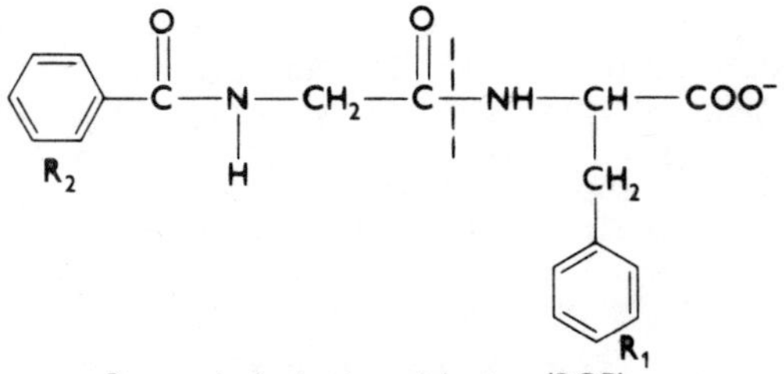

Benzoyl-gly-L-phenylalanine (BGP)⁻ V

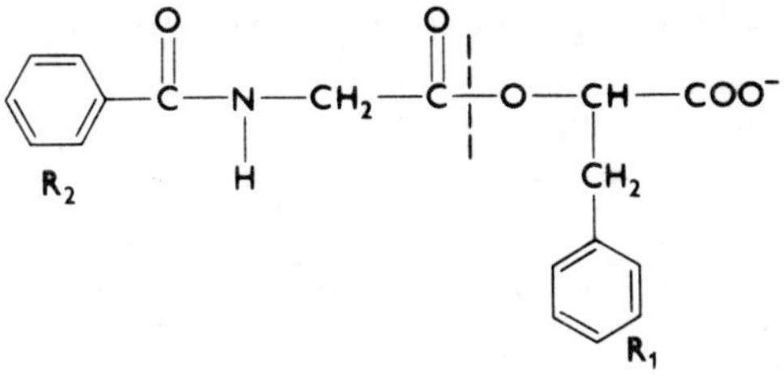

Hippuryl-β-phenyllactate (HPLA) VI

cleaves the C-terminal peptide (or analogous ester) bond of substrates varying from an acyl amino acid to protein molecules. Very extensive early specificity studies established many structural requirements of the substrate which are significant when we discuss the mechanistic features suggested by the X-ray structure of an enzyme-peptide complex.

(1) The C-terminal residue must be of the L-configuration and it must have a free carboxyl. Amides are not hydrolysed, but binding studies show they are bound.

(2) The -NH- of the terminal residue must not be tampered with; substitution on this group severely reduces or abolishes catalysis.

(3) Large aromatic or branched aliphatic side-chains are required on the C-terminal residue for rapid hydrolysis.

(4) Stereochemistry of the penultimate residue is not so critical in that both L-L and D-L peptides are hydrolysed although the D-L peptides are generally less well hydrolyzed.

(5) Activity is greatly enhanced by substitution of the penultimate amino group, especially by an aromatic group such as carbobenzoxy-.

Solution experiments employing radioactive ^{65}Zn showed that the peptide substrates bound to the apoenzyme and prevented Zn binding[12]. This established that the Zn binding site must be closely associated with the substrate binding site. Prevention of Zn binding to the apoenzyme by peptide substrates could also be used to establish the structural requirements for this binding and this method identified three groups as critically involved, the aromatic or branched aliphatic side-chain on the C-terminal residue of L-configuration, the -NH- on the C-terminal residue and the -NH- on the penultimate residue. N-acyl amino acids were not detectably bound to the apoenzyme but required the metal ion for tight binding (see below). The free carboxyl was not required for binding to the apoenzyme, but was required for hydrolysis to occur. Thus the -NH- groups, the C-terminal side chain, and the carboxyl were clear candidates for rather specific interactions with the protein.

Turnover numbers for the hydrolysis of the best peptide substrates by carboxypeptidase A are about 10^2 sec^{-1}. The analogous esters are hydrolysed at about the same rate, but precise comparisons are difficult because of considerable kinetic differences. It is not clear yet whether all features of the mechanism for the two substrates are the same. The esters are differentiated from peptides in their binding behavior in that they do not prevent Zn binding to the apoenzyme. It would appear that tight *specific* binding of the ester requires the metal ion (see ref. 13 for discussion).

Substitution of various first transition metal ions for the native Zn ion has resulted in a series of metallocarboxypeptidases whose catalytic and physicochemical properties form a most interesting chapter in the study of this enzyme. I think these are best discussed after looking at the structure.

One other general area of the solution chemistry that should be discussed before looking at the crystal structure involves the chemical modification of various amino-acid side-chains and the effect of such modifications on catalysis. The most striking of these chemical modifications is with a series of reagents that modify tyrosyl residues. Reagents such as acetic anhydride and acetyl imidazole that acylate tyrosyl hydroxyl groups abolish peptidase activity but enhance esterase activity as much as six-fold. Spectral studies have shown that the modification with acetyl imidazole results in the formation of two O-acetyl tyrosyl residues on the protein. The presence of the competitive inhibitor β-phenylpropionate completely protects against these modifications and leaves activity unchanged after removal of the inhibitor[14]. Thus the prediction was made that at least one tyrosine, possibly two, were intimately involved in the peptidase activity, perhaps not in the esterase activity, and certainly appeared to be near the active site. Much subsequent work has explored these chemical modifications in detail, and by other chemical reagents modified the tyrosyls separately[7], but the conclusion is essentially that given above. Iodination of the tyrosine gives the same results as acetylation and the iodinated tyrosyl residues were isolated in certain specific peptide sequences[15]. This is a clear example of certain organic groups being as essential to the catalysis of peptide hydrolysis as the metal ion. On the other hand it points up what must be some different features of the catalysis of ester compared to peptide hydrolysis, since at least for the substituted ester (HPLA) (VI), the tyrosyl does not appear essential.

Fig. 2. Carboxypeptidase A.

With this brief discussion of the solution enzymology, the general structure of the enzyme as deduced from the 2.0 Å electron density map obtained by Lipscomb and coworkers[8] is shown in Fig. 2.

The general protein structure is of great interest, but can be only briefly commented on here. The enzyme is slightly ellipsoidal, of approximate dimensions 50 Å × 42 Å × 38 Å. About one third, 37.5%, of the molecule is helical, made up of various short segments of helix many of which deviate considerably from the exact folding expected of an α-helix. Several of these are evident on the left of the molecule in Fig. 2. The core of the molecule is made up of several stretches of β-pleated sheet with peptide chains running both parallel and antiparallel. This accounts for 14.6% of the molecule and imparts a very rigid structure to the center of the molecule. The rest of the molecule, e.g. a large part to the right in Fig. 2, can be characterized as random coil, 47.9%. Parts of this random coil region undergo conformational changes on substrate binding. There is one disulfide bond accounting for all the cysteine in the molecule.

The Zn atom is coordinated to the protein by three ligands, His 69, Glu 72, and His 196. The fourth position forming a distorted tetrahedron around the Zn is occupied by a water molecule. The Zn atom is located in a groove and pocket in the enzyme surface, which is lined by the pleated sheet. In the case of carboxypeptidase A speculation on the possible mechanism based on knowledge of the X-ray structure was materially advanced by being able to determine the structure of a glycyl-L-tyrosine complex with the native enzyme. This is a very slowly hydrolysed substrate and is even more slowly hydrolysed if the crystals are cross-linked with glutaraldehyde. By this means a high enough average occupancy at the site of the glycyl-L-tyrosine complex was obtained to determine a structure. The results of this structure are summarized in the schematic Fig. 3A and the structural relations are shown in Fig. 3B.

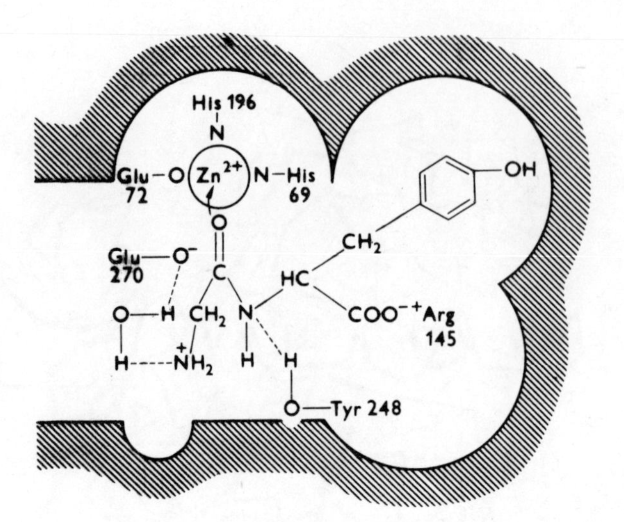

Fig. 3A. Schematic of the carboxypeptidase A-gly-L-tyr complex.

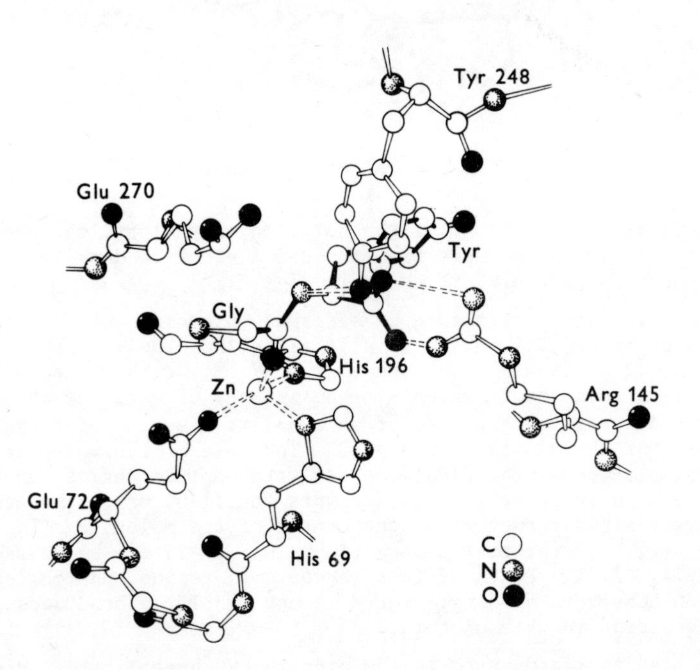

Fig. 3B. Residues at the active site of carboxypeptidase A with gly-L-tyr bound in the active center cavity.

The substrate lies in the groove containing the Zn atom and its ligands with the aromatic side-chain in a pocket that contains no specific binding groups and is large enough to accomodate a tryptophan side-chain. This would appear to account adequately for the observed specificity for the side-chain on the C-terminal amino-acid of the substrate. The free carboxyl group interacts with the positively charged guanidinium group of Arg 145. The carbonyl oxygen of the peptide bond susceptible to hydrolysis replaces the water molecule at the fourth coordination position of the Zn. In an interaction possible only with dipeptide substrates having a free N-terminal amino group,

Glu 270 binds through a water molecule to the free amino group. The oxygen of the hydroxyl group of Tyr 248 is about 2.7 Å away from the nitrogen of the scissile peptide bond and about 3.5 Å from the α-amino group.

The close relations pictured in Fig. 3 between the various atoms of the bound substrate and specific protein groups are brought about by some significant conformational changes in the molecule induced by the binding of glycyl-L-tyrosine. These are summarized in Fig. 4. The guanidinium group of Arg 145 moves 2 Å due to rotation about the C_β-C_γ bond of the side-chain and the carboxylate of Glu 270 also moves by about 2 Å due to rotations about both the C_α-C_β and C_β-C_γ bonds. Most striking of all, Tyr 248 (one of the residues known to be modified in the chemical experiments which so strikingly effect peptidase and esterase activity) moves 12 Å, chiefly by rotation about the C_α-C_β bond to place the -OH in position near the peptide

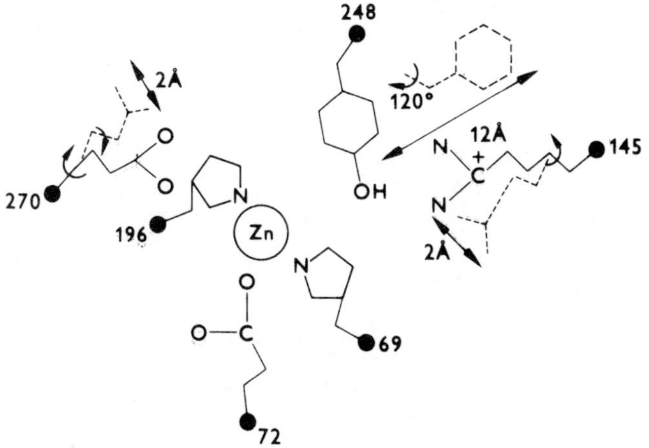

Fig. 4. Conformational changes of active site residues of carboxypeptidase A induced by substrate binding.

nitrogen. There are in addition some apparently coordinated movements of the peptide backbone and a system of hydrogen bonds in the region between Arg 145 and Tyr 248. These conformational changes are an apparent example of the 'induced fit' theory concerning enzyme action proposed by Koshland[16]. This has been widely discussed in terms of enzyme action and proposes that substrate interaction does lead to conformational changes in the protein which are then related to the electronic changes in the enzyme-substrate complex needed to effect catalysis. In carboxypeptidase binding of the C-terminal side-chain in the pocket ejects the H_2O from this cavity and the swinging over of Tyr 248 effectively closes off the active center from the solvent.

Mechanism of Action of Carboxypeptidase A

With this kind of structural information there is at least some structural basis upon which to project the chemical interactions between substrate and protein involved in the mechanism. The apparent binding interactions have been mentioned above, but it should be noted that the interaction between the free carboxylate of the substrate and Arg 145 apparently triggers the changes leading to the rotation of Tyr 248 into the vicinity of the peptide bond. This would clearly explain the absolute necessity for a free carboxyl if the hydroxyl of Tyr 248 near the peptide bond is necessary for peptide hydrolysis.

Thus the coordinated Zn ion, Glu 270, and Tyr 248 are the other groups close enough to the substrate to assist directly in catalysis. Zn can clearly function as a Lewis acid and polarize the carbonyl of the peptide bond. The net charge of the complex is 1+ (2 neutral ligands and a carboxylate), but the single formal net positive charge (this of course does not consider the actual charge distribution at the site which may be more or less concentrated) may be potentiated by the relative hydrophobic environment produced when water is ejected from the side-chain binding cavity and Tyr 248 swings over to partially close the cavity. It can be postulated that the Glu 270 functions in promoting general base attack of the oxygen atom of a water molecule on the carbonyl carbon or itself initiates a specific nucleophilic attack on this carbon. Tyr 248 may then, via a hydrogen bond, donate the proton to the nitrogen as the peptide bond breaks. Another hydrogen bond may be made between the tyrosyl oxygen and the terminal NH_2 (or NH in substituted dipeptides). Such hydrogen bonds could help produce strain at the peptide bond of the substrate. Such postulated mechanistic features as summarized by Lipscomb and coworkers are shown in Fig. 5.

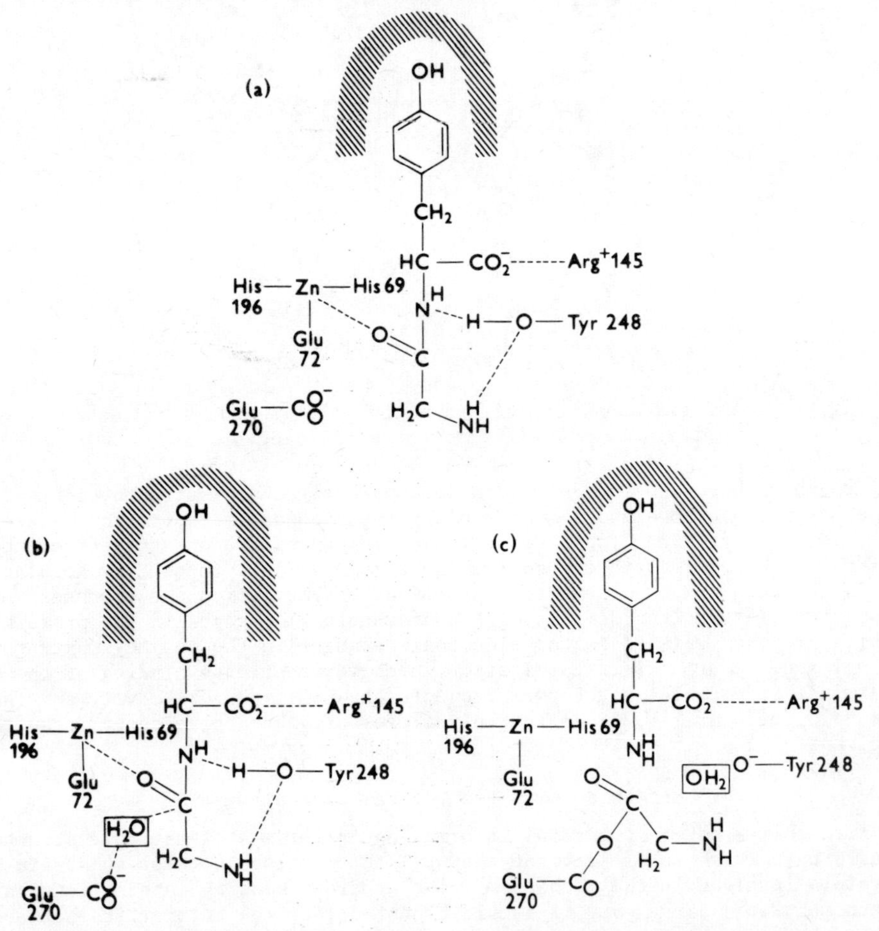

Fig. 5. Possible stages in the hydrolysis of a peptide by CPA. It is probable that the carbonyl carbon of the substrate becomes tetrahedrally bonded as the reaction proceeds, but it is uncertain at what stage of the reaction the proton is added to the NH group of the susceptible peptide bond. (a) Productive binding mode. (b) General base attack by water upon the carbonyl carbon of the substrate. (c) Nucleophilic attack by Glu 270 upon the carbonyl carbon of the substrate.

This would appear to be an example of the metal in a metalloprotein acting much like it is in the model esterase reaction pictured in equation (1), but its polarizing effect is probably assisted significantly by certain features of the nonaqueous protein environment. Most importantly, however, the rapid catalysis would appear to be made possible by a concerted reaction provided by the two protein reactive groups, Glu 270 and Tyr 248, reacting with the two ends of the peptide bond.

The mechanism of ester hydrolysis may involve some similar features, but may differ in a number of respects. At least for hippuryl-β-phenyllactate, the chemical modification studies suggest that the Tyr 248 interaction is not necessary for hydrolysis. Tyr 248 has recently been modified with an arsanilazo group and studies with arsanilazo carboxypeptidase in both solution and the crystalline state are providing more information about the role of this group in catalysis[17].

METAL ION SPECIFICITY

It is possible in many metalloenzymes to substitute the native metal ion with another first transition or IIB ion either by removing the original ion with chelating agents or by exchange dialysis[18,19]. In addition to certain interesting physical properties that can be studied by this device (see below), it provides an opportunity to discover what kind of specificity for catalysis resides in the electronic properties of the metal ion. This has now been done for several metalloenzymes, most extensively for carboxypeptidase A[7,18,19], and the results are grouped together in this section.

In model metal-catalysed reactions like the ester hydrolysis represented in equation (1), catalysis is initiated by complex formation with the metal and as expected many of these show an increased or decreased rate with different metal ions depending on the relative formation constants characterizing the complex between the substrate group and these metal ions. For the commonly employed first transition and group IIB metal ions the series goes roughly as follows:

$$Mn(II) < Fe(II) < Co(II) < Ni(II) < Cu(II) > Zn(II) \geqslant Cd(II) < Hg(II).$$

Several at least of these metals have been tried as activators of the apoenzymes produced from several naturally occurring Zn(II) metalloenzymes and the results are given in Table III. Metals successful in inducing significant enzymatic activity are listed in order of decreasing activity.

TABLE III

Metal Ion Specificity for Catalysis by Several Metalloenzymes

Enzyme		Active metals	Inactive metals
Carboxypeptidase A	(Peptidase)	Co > Zn > Ni > Mn > Fe	Cd, Hg, Cu
	(Esterase)	Cd > Hg > Mn > Co > Zn > Ni	Cu
Carbonic anhydrase	(CO_2 hydration)	Zn > Co	Mn, Ni, Cu, Cd, Hg
	(Esterase)	Co > Zn	Mn, Ni, Cu, Cd, Hg
Alkaline phosphatase	(Ester hydrolysis)	Zn > Co	Mn, Ni, Cu, Cd, Hg
	(Formation of phosphoryl enzyme)	Cd > Zn > Mn > Co ?Cu	Ni, Hg
Alcohol dehydrogenase		Zn > Co > Cd	

In all cases listed in Table III, the evidence is good that all the metal ions listed for the same enzyme occupy the same site as the native Zn(II) ion, although the methods thus far would not distinguish altered coordination geometries at the site.

The most general characteristic evident from the data in Table III is that efficient catalysis depends rather selectively on which metal ion is present, and the relative activity of the metal ions that do induce activity bears no clear relation to the relative stability of the complexes they form in solution. Cu(II) which forms the most stable complexes in solution and is a commonly used catalytically effective metal in model systems (e.g. the two ester hydrolyses mentioned earlier, equation (1), I and II) is uniformly inactive in these enzymes. Since copper prefers the square-planar distortion due to the Jahn-Teller effect occuring in its complexes, the induction of this distortion at the active sites which are not planar in their active form is often invoked to explain this loss of activity on Cu(II substitution. While there is some evidence from ESR studies that this may b true, there is no direct information as yet as to whether the protein ligand are flexible enough to allow this.

Efficient catalysis is also rather selective in at least two of the cases, carbonic anhydrase and alkaline phosphatase, being limited to Co(II) and Zn(II). The reasons for this particular pair of metal ions being the most effective catalysts at these protein sites is not readily explained on the basis of present comparative coordination chemistry.

A brief comment needs to be made about 'efficient' and 'non-efficient' catalysis in these systems. Zn(II) levels can only be reduced by the best methods to somewhere around 10 n mol/l. It is almost impossible to remove the last traces of Zn(II) from many of these enzymes. Since enzymes are active at such small concentrations there is always at best a 1-5% backgroun activity arising from Zn(II) contamination. Thus the only meaningful difference at the very best is between 1 and 100% activity or two orders of magnitude. Thus we can say that an inactive metal is at least 10^2 less effective but we cannot eliminate the possibility that this might actually be 10^{10} less effective. Such differences would be very significant.

Carboxypeptidase A might be said to be more like model systems, since more metals are active. Closer inspection of the system, however, reveals some most unusual characteristics. The order given in Table III for peptide hydrolysis applies to the most commonly used substrate, carbobenzoxy-glycyl-L-phenylalanine, with Co(II) carboxypeptidase the most active. This order, however, depends upon which peptide is used, a phenomenon which is not readily explained. Even more surprising the Group IIB metal ions, Cd(II) and Hg(II), have no peptidase activity, but are more active than Zn(II) as esterases. The origins of these effects are not yet clear. It has been tempting to relate the phenomena with Cd(II) and Hg(II) to their larger ionic radii and there are some X-ray data suggesting that Hg(II) distorts the active site complex, possibly sufficiently to prevent the interaction of Tyr 248, which as chemical modification shows may not be involved in HPLA hydrolysis. At present highly refined X-ray data on the different metal derivatives are not available. It is possible that high-resolution three-dimensional structures will reveal significant coordination differences among the different metal derivatives. Most data at present are obtained by way of Fourier difference maps which reveal only the approximate coordinates, with no details.

CARBONIC ANHYDRASE

Among enzymes catalysing hydration-dehydration reactions, carbonic anhydrase has been of particular interest to biochemists and physiologists. partly because it catalyses an important physiological reaction, the hydration of CO_2, which occurs rapidly in the absence of catalyst. The rate enhancement by carbonic anhydrase, however, is particularly striking: the first-order rate constant for the catalysed reaction at pH 7 is near 10^5 sec^{-1}, making this one of the fastest enzyme reactions known. The first-order rate constant for the uncatalysed reaction is $\approx 3\times10^{-2}$ sec^{-1} at 25 °C. Carbonic anhydrase was also the first zinc metalloenzyme to be discovered, containing 1 mole of zinc per ≈30 000 daltons[20]. Since the early work involving inhibition with metal complexing agents, it has been amply proved that the metal is essential for catalysis[21,22] (for review[22a]). One interesting aspect of carbonic anhydrase which I shall not cover is the distribution and multiple isozyme nature of the enzyme. The enzyme occurs in many animal and plant tissues where the rapid hydration of CO_2 or dehydration of HCO_3^- (H_2CO_3) is required for various ion-transport processes. The most accessible location is the red blood cell where it occurs in high concentrations. Some animals including man possess two distinct isozymes differing in amino-acid composition and hence primary structure (perhaps due to gene duplication). These isozymes do not have identical enzymatic properties, but the general properties discussed below will not differ significantly as a function of isozyme.

The catalysed reaction can be formulated in two ways; one with

$$CO_2 + H_2O \rightleftharpoons H_2CO_3 \qquad (2)$$

$$CO_2 + OH^- \rightleftharpoons HCO_3^- \qquad (3)$$

carbonic acid as the substrate in the dehydration reaction (equation (2)) and one with the hydrogen carbonate ion as the substrate in the dehydration reaction (equation (3)). The two are related of course by the equilibrium for the dissociation of carbonic acid. The choice of which species is the actual substrate for the enzyme, however, has different kinetic consequences as will be discussed below. The maximum rates of hydration and dehydration as a function of pH have opposite pH-rate profiles. The maximum velocity of hydration varies as if it depended on the basic form of a group on the protein with a pK_a between 7 and 8, while the maximum velocity of dehydration varies with pH as if it depended on the acid form of this group. Any proposed detailed mechanism will have to follow the general rate laws which describe the inverse pH-rate profiles[23].

$$v_h = k_h\,[CO_2]\,[E] \qquad (4)$$

$$v_d = k_d\,[H^+]\,[HCO_3^-]\,[E] \qquad (5)$$

E must be the same form of the enzyme in both equations, k_h and k_d are the rate constants for hydration and dehydration respectively.

In recent years some additional reactions involving carbonyl groups, hydration and hydrolysis, have been discovered which probably do not have physiological significance, but have been useful in studying physicochemical features of the enzyme. These are listed below with the relevant references, but they will not be discussed in detail. They include the hydrolysis of a number of esters, represented by *p*-nitrophenyl acetate and a cyclic sultone (equations (6,7)), and the hydration of aldehydes represented by acetaldehyde (equation (8)).

Sulfonamides (VII) were early discovered to be highly specific and potent

$$R—SO_2—NH_2 \qquad (VII)$$

inhibitors of this enzyme along with metal binding anions such as CN^-, HS^-,

OCN^-, and N_3^-. While these strongly metal-binding anions are prominent members of the class, a great variety of anions appear to bind including Cl^-, Br^-, I^- and compounds with carboxylate groups. Much of the solution chemistry suggests that the anion interaction as well as that of the sulfonamides involves direct coordination to the metal ion. This conclusion has now been confirmed by the X-ray studies. The solution studies have been reviewed extensively[10,21,22,22a] and I will mention certain results as they apply to the structure.

$$CH_3-\overset{O}{\overset{\|}{C}}-O-C_6H_4-NO_2 + H_2O \rightleftharpoons CH_3COO^- + {}^-O-C_6H_4-NO_2 + 2H^+ \quad (6)$$

(ref. 24,25)

$$\text{2-hydroxy-5-nitro-}\alpha\text{-toluenesulfonic acid sultone} + H_2O \rightleftharpoons NO_2C_6H_3(OH)CH_2-SO_3^- + H^+ \quad (7)$$

(ref. 26)

$$CH_3-\overset{O}{\overset{\|}{C}}-H + H_2O \rightleftharpoons CH_3-C(OH)_2-H \quad (8)$$

(ref. 27)

One area of the solution chemistry is given here, as an example of a general approach that has proved valuable in a number of these Zn metalloenzymes and is potentially valuable in the study of most metalloenzymes. The interesting effects of metal substitution on enzymatic activity were discussed in the previous Section, but if the substitution involves a transition metal ion, then the possibility arises of finding a good deal out about the metal ion environment by looking at the optical spectra arising from the *d-d* transitions and in favorable cases the ESR transitions. Both approaches have been valuable in the study of metal-substituted carbonic anhydrases and the results will serve here as brief examples of what should be a quite general approach.

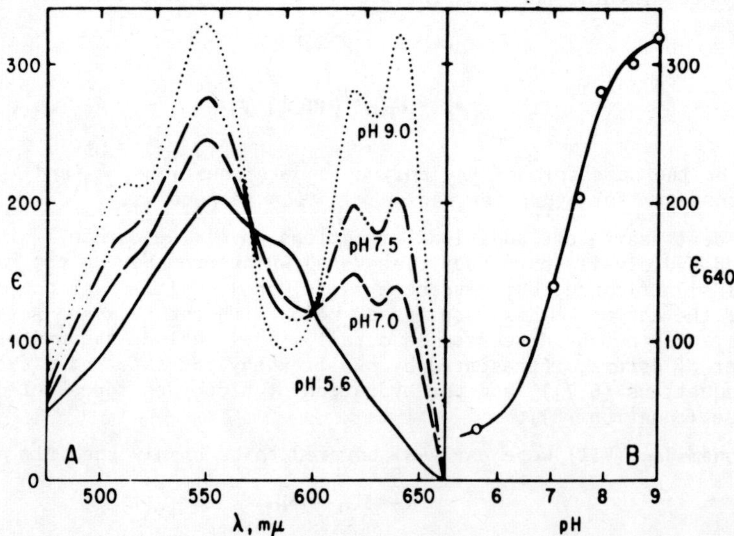

Fig. 6. Visible absorption spectrum of Co(II) carbonic anhydrase as a function of pH.

The *d-d* transitions of Co(II) carbonic anhydrase are located between 450 and 650 nm as shown in Fig. 6 from Lindskog and Nyman[28]. The general intensity of the spectrum is that expected of a tetrahedral complex of Co(II) although the band structure at high-pH implies a distorted complex. The most interesting feature is that the spectrum changes from one species at pH 6 to another at pH 9. This change follows a sigmoid titration curve with a pK_a between 7.5 and 8. A large amount of investigation has shown that this appears to be the same pK_a reflected in the hydration-dehydration pH-rate profiles. This is among the strongest evidence that the pK_a is very closely associated with the metal ion itself, which is discussed below; but the favored postulate at present is that this pK_a represents that of a metal-coordinated water molecule.

The second example of the spectral approach is shown in Fig. 7 by the ESR spectrum of a CN^- complex of $^{63}Cu(II)$-substituted carbonic anhydrase. Among

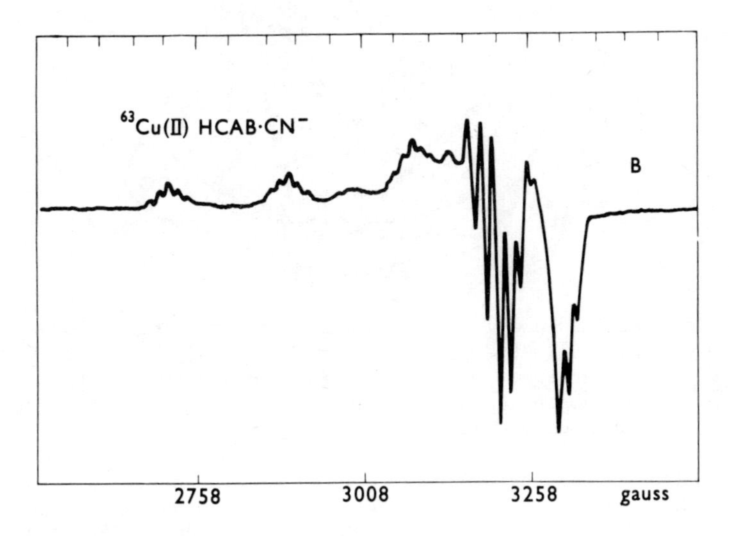

Fig. 7. ESR spectrum of Cu(II) carbonic anhydrase-cyanide complex.

the various conclusions about the molecular structure at the active site that can be obtained from this spectrum is that there are two magnetically equivalent nitrogen nuclei as ligands to the copper ion. There are five nitrogen superhyperfine lines on each copper hyperfine line. These two examples illustrate the potentialities of these methods.

As in the case of carboxypeptidase, the possible mechanism of catalysis is discussed best after looking at the X-ray structure. In this case, however, it is perhaps less helpful in speculations about mechanism because the nature of the reaction makes it extremely difficult to devise a means of looking at an 'enzyme-substrate' complex.

The general results of the X-ray structure at 2.0 Å resolution are shown in Fig. 8. The molecule is nearly spherical with dimension approximately 41 Å × 41 Å × 47 Å measured between the extreme points of the backbone. The Zn atom is located near the center of the molecule coordinated to three amino-acid side-chains of the protein, His 93, His 95, and His 117. A fourth coordination position is occupied by solvent H_2O or OH^- to complete a considerably distorted tetrahedral geometry around the Zn(II). The first two ligands have been positively identified in the sequence. His 117 has not been positively identified as yet in the sequence. Why the Cu(II) ESR signal (Fig. 7) shows only two equivalent nitrogens in the cyanide complex must provide some clue to structure at the active center if the third ligand is definitely histidine. If there are three histidyl side-chains as ligands, it would appear that the Zn center has a formal 2^+ positive charge with three neutral ligands unless there is some unusual pK for one of the ring protons.

Other general features of the molecule include a broad twisted sheet of parallel and antiparallel β-pleated sheet strands running through the molecu This is represented by the broad arrows in Fig. 8, and is rather similar to the rigid structure running through the carboxypeptidase A molecule. This

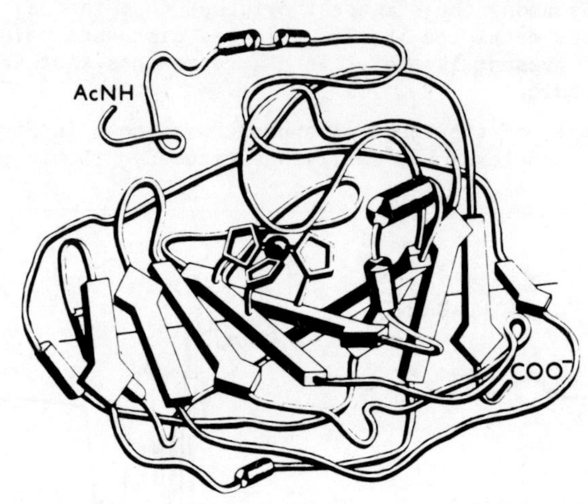

Fig. 8. An idealized drawing of the main chain folding of human carbonic anhydrase C. The helices are represented by cylinders and the pleated sheet strands are drawn as arrows in the direction from amino to carboxyl end. The ball supported on three histidyl residues is to be identified with the Zn ion associated with the active site.

represents about 37% of the molecule. A few widely scattered sections of helix (cylinders in Fig. 8) account for 20% of the remaining residues. Very few residues in these helical segments actually belong to the classical Pauling α-helix. The three ligands to the Zn(II) ion all come from the center section of the pleated sheet structure and place the Zn ion at about the center of the molecule at the bottom of a funnel-shaped cavity ≈15 Å deep. The upper part of the molecule (Fig. 8) forms part of this cavity and the β-structure forms the bottom and part of the funnel-shaped mouth of the cavity. The lining of the cavity contains some non-polar residues, but also a significant number of polar residues including two histidyl residues, His 63 and His 128, besides those binding the Zn. The funnel-shaped cavity with the Zn(II) ion at the bottom clearly represents the active center of the molecule.

A detailed summary of the structural relations of some of the residues in this cavity is illustrated in Fig. 9. Only the three histidyls and the water molecule are close enough to interact directly with the metal ion. Others like His 63, Glx 66, His 128, and Leu 196 are 4 to 5 Å away. Another interesting feature of the cavity is that it is filled with highly structured or 'ice-like' water molecules. This structured solvent is illustrated in Fig. 10, the open circles showing the solvent molecules and the lines a network of possible hydrogen bonds.

When an inhibitor like a sulfonamide is bound, it binds in the cavity and displaces all the organized solvent (Fig. 11). The sulfonamide group (mercurated in Fig. 11) itself binds to the Zn(II) ion, probably through coordination of the nitrogen (but oxygen is not ruled out). This is in agree-

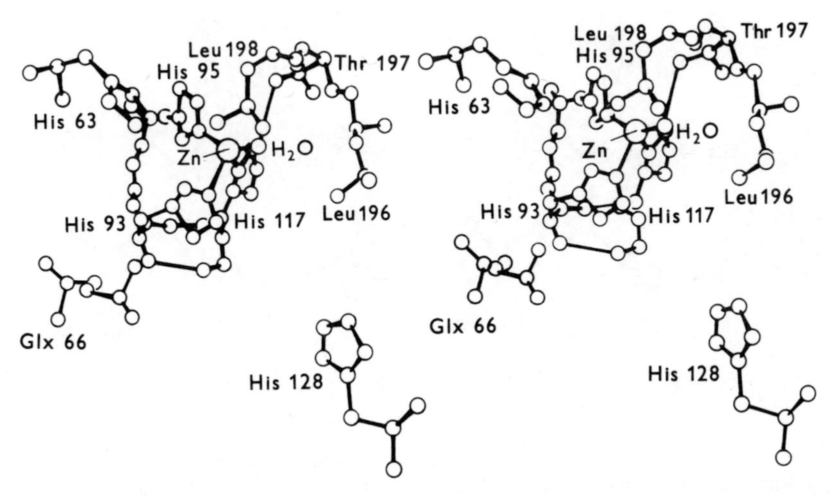

Fig. 9. Stereoscopic drawing of some residues surrounding the Zn ion as well as two histidyl residues further out in the active site of human carbonic anhydrase C.

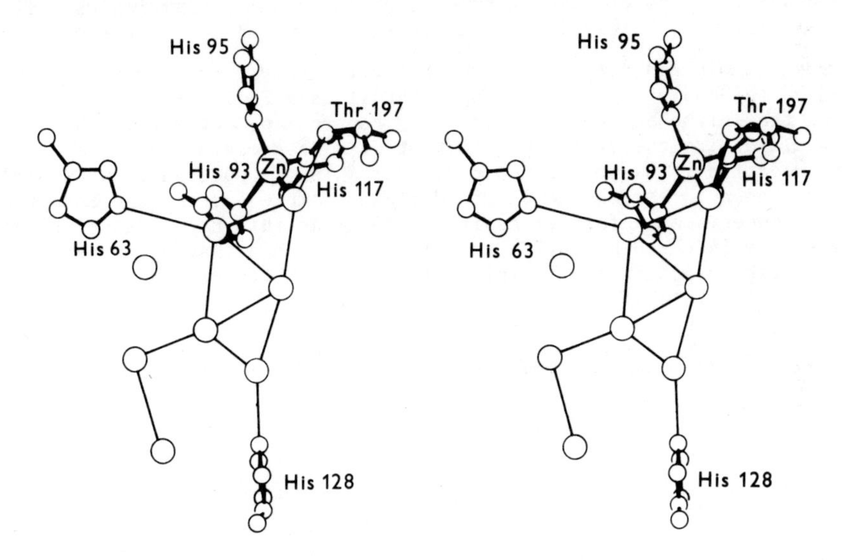

Fig. 10. Structural water (open circles) in the active site cavity of human carbonic anhydrase C.

ment with earlier solution data showing that the metal is necessary for sulfonamide binding. Spectral data also show that the bound form of the sulfonamide is the anion $-NH^-$ in complete analogy to formation of a simple metal-ligand complex with a protonatable ligand. The H_2O or OH^- in the fourth coordination position is displaced. Other anion inhibitors also appear to occupy the coordination position normally occupied by solvent, but the data are not so complete as yet. Solution data already show that at high pH the binding of an anion ligand like CN^- is accompanied by H^+ uptake or OH^- release while at low pH the binding of a species like HCN is accompanied by release of an H^+.

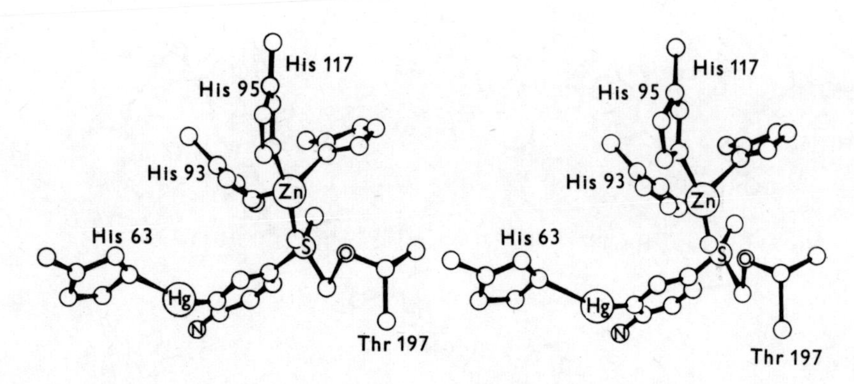

Fig. 11. Stereoscopic drawing of the AMSulf molecule binding at the active site of human carbonic anhydrase C.

Since the enzyme catalyses such a simple reaction (albeit extraordinarily rapidly) and we have this much structural information on the molecule, one would think that selection of the most reasonable mechanism of catalysis would be relatively easy. It has not proved to be so. The substrates are small and even if a crystal could be adjusted to contain significant equilibrium concentrations of some intermediate, the density differences will be small. It may be possible in the future to accomplish this, however.

The most simple and elegant mechanism for the enzyme could already be postulated on the basis of the knowledge that it is a Zn(II) enzyme and that a group on the enzyme between pH 7 and 8 is controlling activity. This mechanism would assign the latter pK_a to that of a Zn-coordinated water molecule. This is relatively low for such a pK_a, but pK_a's of coordinated water molecules are known to be lowered in mixed ligand complexes, and the enzyme microenvironment could further assist in lowering this pK_a. Assuming that the substrate in the dehydration reaction is HCO_3^- and that all substrate reactions take place in the vicinity of the Zn-H_2O center, then the postulated mechanism is as follows:

$$\mathrm{EnZn{-}OH^- + CO_2} \underset{k_{-1}}{\overset{k_1}{\rightleftharpoons}} \mathrm{EnZn \cdot HCO_3^- + H_2O} \underset{k_{-2}}{\overset{k_2}{\rightleftharpoons}} \mathrm{EnZn \cdot OH_2 + HCO_3^-} \qquad (9)$$

$$\mathrm{EnZn \cdot OH_2 + HCO_3^-} \underset{k_{-3}}{\overset{k_3}{\rightleftharpoons}} \mathrm{EnZn{-}OH^- + H^+ + HCO_3^-}$$

This simple mechanism accounts for a remarkable number of facts about the enzyme. Zinc is required to make the attack of OH^- on CO_2 a significant reaction at neutral pH ensuring that OH^- is the reactive species at the active site. This reaction would not make a significant contribution to the uncatalysed reaction until above pH 10 where it actually goes quite rapidly. Since according to equation (9) the hydration rate is proportional to $[CO_2]$ and the dehydration rate is proportional to the product $[H^+][HCO_3^-]$, the mechanism follows the required inverse pH-rate profiles. Data on proton equilibrium accompanying inhibitor binding also receives a simple explanation in terms of known coordination chemistry. Binding of the neutral species of inhibitors, e.g. HCN, H_2S, or R-SO_2-NH_2 at low pH will be accompanied by H^+ release, binding of CN^-, HS^-, or -NH^-at high pH will be accompanied by OH^- release. This is entirely as expected from known coordination chemistry. Although the anionic species could be considered as inducing H^+ uptake by binding at high pH, mechanisms for this then require some more complex postulates like alterations in pK_a of the anion or adjacent protein groups. As far as support for this mechanism from the X-ray structure is concerned, the presence of a solvent-occupied coordination position is compatible. Also there are no other groups in the immediate vicinity of the metal ion that can participate directly in catalysis except possibly in substrate binding or possibly by formation of certain hydrogen bonds. Several of the non-Zn-coordinated histidyl residues have been chemically modified with some effects on enzyme activity, but no basic alteration of mechanism[30].

Mechanisms involving H_2CO_3 as substrate have been proposed incorporating the $ZnOH_2 \rightleftharpoons ZnOH^+ + H^+$ equilibrium[10]. If the ionization on the protein participating in activity is not assigned to the $ZnOH_2$, then another group participating in the mechanism, or a composite pH-change in the enzyme producing an apparent pK_a must be evoked. At present it is difficult to visualize what this may be. Based on the X-ray structure and the chemical modification data no likely candidate for an additional group near the Zn atom participating directly in catalysis suggests itself.

Lastly there are some kinetic difficulties with all the proposed mechanisms when the extremely rapid turnover number, 10^5 sec^{-1}, of carbonic anhydrase is considered. In the mechanism considered in equation (9), the limitation on the rate of diffusion-controlled proton transfer suggests that, at least in model systems, the proton could not be transferred fast enough in step 3 to account for the rate. In mechanisms involving H_2CO_3 as substrate, the concentrations at pH 7 would be so low that both diffusion problems and the rate of protonation of HCO_3^- to form the substrate present rate difficulties, at least when based on numbers from model systems. These are unsolved problems in explaining how carbonic anhydrase catalyses the reaction so rapidly.

Crystals of Cu-, Co-, Mn-, and Hg-carbonic anhydrase have been examined by X-ray diffraction producing electron-density difference maps. All the metal ions except Hg(II) occupy the same site as Zn(II). Resolution is not good enough yet, however, to exclude small differences in the coordination geometry. The Hg^{2+} ion is displaced about 0.6 Å from the position of the others, perhaps due to distortions caused by the large ion. Only a high-resolution structure of each of these derivatives will give the possibility of detecting differences that might suggest why Zn(II) and Co(II) are active while Mn(II), Cu(II), Cd(II), and Hg(II) have low or no activity.

ALKALINE PHOSPHATASE

Alkaline phosphatase is a protein dimer of mol. wt. ≈80 000 containing two Zn(II) ions per dimer which are essential to activity[10] (see ref. 29 for review). The dimer can bind more Zn(II) ions (up to 10), but their relation to activity is unclear at the present time. Evidence that the active protein is made up of two identical 40 000 mol. wt. monomers comes from several lines of evidence: there is only one structural gene for the enzyme, it reversibly dissociates at pH 2 into 40 000 units which reactivate at pH 7 where the hydrodynamic species is 80 000, and the crystal structure shows two tightly associated monomers related by a two-fold axis.

This enzyme is used here as an example of a metalloenzyme involved in a highly important area of biological chemistry, that of phosphate ester hydrolysis and phosphate group transfer. It also illustrates the increased complexity that occurs when a multimeric protein is involved in catalysis. Most of the chemistry has been done on the inducible enzyme from *Escherichia coli*; however, much evidence indicates that the mammalian alkaline phosphatases share many of the physicochemical properties including a catalytically essential Zn(II) ion.

The physiological function of the enzyme is not well understood, but human mutants who have defective bone alkaline phosphatase die at an early age from failure to calcify their bones.

The reaction catalysed by the enzyme can be formulated as in equation (10).

$$E + ROP \rightleftharpoons E \cdot ROP \underset{k_{-2}}{\overset{k_2}{\rightleftharpoons}} E{-}P \underset{k_{-3}}{\overset{k_3}{\rightleftharpoons}} E \cdot P \rightleftharpoons E + P \qquad (10)$$

$$+ \; R'OH \rightleftharpoons R'{-}OP + E$$

E-P refers to a phosphoryl enzyme believed to be a phosphoserine based on the finding that at pH 5.5, where the observation of 'burst' kinetics shows that dephosphorylation of E-P, k_3, is rate limiting, a peptide containing Ser-O-P can be isolated. This is produced either by the hydrolysis of R-O-^{32}P or incubation with ^{32}P inorganic phosphate, both compatible with equation (10). The sequence of the ^{32}P-labelled peptide is Thr-Gly-Lys-Pro-Asp-Tyr-Val-Thr-Asp-Ser^{32}P-Ala-Ala-Ser-Ala. Although the enzyme is maximally active at pH 8-9, a region where this phosphoserine cannot be isolated, the low pH finding suggests strongly that the Ser-O-P is an important step in the mechanism. As the reaction gets very fast at alkaline pH, the equilibrium concentration of this intermediate must get very small. The reason for this is actively under discussion[29].

Zinc has been shown to be essential to induce the binding of phosphate, essential for the formation of the phosphoryl enzyme, and also catalyses its breakdown[30]. Before discussing additional solution chemistry, the structural information available from X-ray crystallography will serve to focus discussion. Unfortunately the crystallography of this enzyme has not progressed to high resolution at this time and the information available from a low-resolution map, 7.7 Å, is presented in Fig. 12, a balsa model of this map. The molecule has been cut at the level of the two Zn(II) ions and shows two very tightly interacting monomers (no cleavage plane discernible at this resolution) arranged around the two-fold axis in the center[31]. Each of the Zn(II) ions is located 16 Å from the diad axis. When Hg(II) is added to the apoenzyme produced by removal of the Zn(II) from the crystal with chelating agents, the Hg(II) occupies the Zn(II) sites.

Difference maps between the Zn(II) enzyme and the apoenzyme show considerable electron-density redistribution in an area about 12 Å from the diad axis. This may be interpreted as a fairly large conformational change, perhaps at the monomer-monomer interface when Zn(II) is removed. Any exact interpretation will require higher resolution. Other features of the molecule are not resolved at 7.7 Å.

With this structural information available some of the solution data suggest additional complexities in the mechanism of catalysis. Phosphate binding studies, ^{32}P-labelling studies, and 'burst kinetics' at low pH have all

Fig. 12. Model of alkaline phosphatase cut through the plane of the Zn ions based on X-ray data at 7.7 Å resolution.

shown that only one active site can be labelled or is operating at any instant, even though the enzyme must contain two identical metal sites related by two-fold symmetry. The electron-density map clearly shows that the two Zn(II) ions are too far apart for both to be located in a single active site. Thus the stoichiometry of substrate interaction suggests that there must be a high degree of negative cooperativity between the monomers, such that once a substrate binds in one site the other site can no longer bind substrate. Negative cooperativity could result from conformational changes induced by substrate binding which are propagated between monomers, perhaps even destroying the precise two-fold symmetry. Positive cooperativity of this sort has been well documented in the case of hemoglobin[32]. When Cu(II) occupies the Zn(II) sites, the nuclear superhyperfine structure on the ESR signal at pH 8 identifies three magnetically equivalent nitrogen atoms as ligands to the Cu(II)[33]. The most likely candidates for these ligands are the ring nitrogens of histidyl residues.

The idea of two identical active sites which interact via negative homotropic interactions as outlined above assumes at least as a working hypothesis that the metal is contained within the locus that binds the substrate. The evidence for this, however, is not nearly so great as in the monomer systems just discussed. The metal ion is clearly, however, involved in the various catalytic steps and the indirect evidence is briefly listed below.

(1) A Co(II) enzyme is active and shows a visible absorption spectrum arising from two Co(II) ions much like that of Co(II)-carbonic anhydrase (Fig. 6). Phosphate alters this spectrum, and one phosphate group alters the spectrum of both Co(II) ions.

(2) Formation of the phosphoryl enzyme is metal-catalysed; the apoenzyme will not form a phosphoryl enzyme.

(3) The rate constants characterizing phosphorylation (k_{-3}) and dephosphorylation (k_3) from inorganic phosphate depend on the species of metal ion (Table III). Zn(II) catalyses both fairly rapidly. Cd(II) slows both, but dephosphorylation by Cd(II) is so slow at all pH values that the enzyme is essentially inactive. Mn(II) is similar. Co(II) is like Zn(II). That the metal is involved in catalysing dephosphorylation of the serine is confirmed by preparing an apophosphoryl enzyme by phosphorylating the Cd(II) enzyme and then removing the Cd(II). This phosphorylated apoenzyme is stable at all pH values, but is rapidly dephosphorylated by adding Zn(II).

(4) With high enough concentrations of a second alcohol (R' equation (10)) the enzyme will transfer the phosphate group from E-P to form R'-OP. The acceptor OH must be located on a carbon adjacent to one carrying an amino group. Hence ethanolamine and tris(hydroxymethyl) methylamine are good acceptors. The species of metal greatly affects the efficiency of this transfer reaction, e.g. Zn(II) catalyses this transfer, Co(II) does not, at alkaline pH.

At the present stage of knowledge, possible mechanisms are unclear. One readily suggests itself, indicated in VIII, but enzymes tend to make liars out of postulating biochemists. However, assuming that the metal interacts

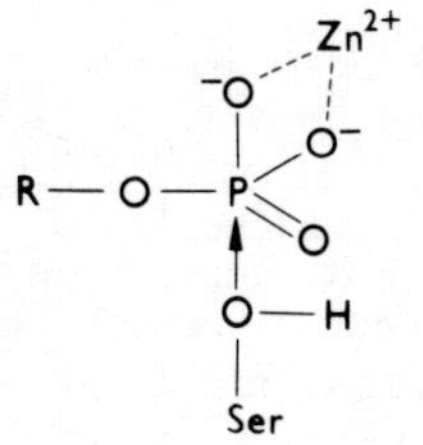

VIII

directly with the phosphate (for which there is no direct evidence), a simple mechanism would have the Zn(II) forming a complex with the negative oxygens, thereby reducing the charge density on the phosphorus. This would then favor a nucleophilic attack of the oxygen of the serine OH on the phosphorus, a mechanism that is usually not favored in phosphate ester hydrolyses. Other mechanisms are possible (a metaphosphate mechanism is not totally ruled out) and possible rearrangements of a five-coordinate intermediate are interesting aspects of the phosphatase mechanism.

FERREDOXINS

Metalloenzymes are exceedingly numerous among the class of enzymes that catalyse oxidation-reduction reactions. A very brief review of the breadth of this area is provided by the limited examples of metallo-oxido-reductases listed in Table I. The field is so broad and involves so much complex chemistry that almost any brief discussion can be considered as weighted in favor of one group or another of these enzymes, for example heme proteins. Instead of trying to be inclusive in this area I have limited this presentation to one class of these proteins, the small non-heme iron proteins known as the ferredoxins (Fd). The selection is somewhat arbitrary, but they do have some favorable characteristics which facilitate discussion at the present time and which I hope will justify using a specific example as an introduction to this complex area of metalloenzyme catalysis. (1) They are small proteins, mol. wt. = 6000 - 25 000, hence many of the primary structures were relatively easy to obtain. (2) They contain only iron and inorganic sulfide in addition to the amino acid chain, hence we are dealing with an Fe-protein interaction without the additional complexity of a prosthetic group as in the heme proteins. (3) These iron-sulfur proteins are the object of intense current investigations so that detailed information including the X-ray structure is now available for several of them. (4) They are involved in a variety of interesting biological oxidation-reduction reactions related to photosynthesis, anaerobic bacterial metabolism, and steroid biosysthesis. The same type of iron-sulfur center is probably present in mitochondria and is involved in electron transport in the respiratory chain of enzymes. Several examples of the ferredoxin class of enzymes are given in Table IV. One of the most interesting chemical properties is the range of oxidation-reduction potentials observed for these proteins at neutral pH, over 700 mV, from -400 mV for the bacterial and plant ferredoxins to +350 mV for the high-potential protein isolated from *Chromatium* (Table IV).

TABLE IV

Oxidation-Reduction Potentials for Ferredoxins

Enzyme	Source	Mol. wt.	Ox. red. pot. (V)	Iron	Sulfide	Cysteine residues
Ferredoxin	Spinach	13 000	-0.432	2	2	-
Ferredoxin	*Clostridium pasteurianum*	6 000	-0.418	8	8	8
	Micrococcus aerogenes	6 000		8	8	8
Adrenodoxin	Adrenal cortex	22 000	-0.270	2	2	-
Rubredoxin	*Clostridium pasteurianum*	6 000	-0.057	1	0	4
High-potential iron protein (HiPIP)	*Chromatium vinosum*	9 500	+0.350	4	4	4

This remarkable range of oxidation-reduction potentials is presumably required by the immediate substrates that reduce or reoxidize these iron proteins. The widely differing oxidation-reduction potentials must ultimately be related to small differences in structure among this group of proteins with similar general structure. Just how similar will be indicated by the X-ray studies below.

Before looking at the physicochemistry of the isolated molecule, a brief resumé of the function of these molecules is appropriate. The subject has been extensively covered in a number of reviews[34-36] and is rather complex. Only the briefest outline is appropriate here.

Green Plants

Proteins of the ferredoxin-type were first isolated from green plants and it is now believed that ferredoxin functions as the first electron acceptor for the photoactivated chlorophyll molecule (equation (11)). The reducing

$$\begin{array}{ccc} NADP \underset{}{\overset{e^-}{\rightleftharpoons}} & Ferredoxin & \\ & \uparrow & \rfloor\lceil ADP \; P_i \\ ATP \xleftarrow{ADP + P_i} & Chlorophyll & \lfloor ATP \rightarrow O_2 \\ & \uparrow & \leftarrow H_2O \\ & h\nu & \end{array} \qquad (11)$$

electrons originate from water, pass to chlorophyll and then to ferredoxin. Reduction of ferredoxin is closely coupled to the evolution of oxygen and in the presence of ADP and inorganic phosphate is coupled to the formation of the high-energy phosphate of ATP in the process known as non-cyclic photophosphorylation (left, equation (11)). For each mole of oxygen evolved, four ferredoxins are reduced, and two ATP's are formed (equation (12)).

$$4H_2O + 4Fd_{ox} + 2ADP + 2P_i \xrightarrow{h\nu} O_2 + 4Fd_{red} + 2ATP \qquad (12)$$

The observed stoichiometry agrees with the observation that reduction of ferredoxin involves a 1-electron transfer; one ATP is formed for each pair of electrons transferred from water to ferredoxin.

In this system, ferredoxin's oxidation-reduction potential of -432 mV is 100 mV lower than that of the $NAD(P)^+$-NAD(P)H couple, hence Fd can reduce the pyridine nucleotide coenzymes and can function as the electron acceptor between photoactivated cholorophyll and the enzymes reducing NADP. The latter coenzyme molecules are the reducing equivalents used in the synthetic reactions of photosynthesis.

If there is no net electron flow and no water is consumed and no oxygen is evolved, the electrons from the reduced ferredoxin are transported back to the chlorophyll in a series of reactions the details of which have not been elucidated. They apparently bear some similarity to the coupled oxidative-phosphorylation system present in mitochondria, since the energy is coupled to the production of ATP from ADP and inorganic phosphate (right, equation (11)), the so-called cyclic photophosphorylation.

Anaerobic Bacteria

In anaerobic bacteria (e.g. the *Clostridia* from which many of the ferredoxins have been isolated) the anaerobic oxidation of certain substrates is coupled to the production of ATP and a reductant in the process of fermentation. ATP and the reductant are both required for synthesis of new cell substances. These reactions are complex and mechanisms are incompletely studied in many cases except that ferredoxin has been shown to be required as the reductant.

Anaerobic bacteria catalyse the synthesis of acetyl coenzyme A from pyruvate, a reaction also central to the intermediary metabolism of mammalian cells. The anaerobic reaction was known to be different from the one in aerobic cells, since there was no requirement for lipoic acid or NAD. In addition the overall reaction was associated with the production of hydrogen gas (equation (13))

$$\text{Pyruvate} + \text{CoA} \rightleftharpoons \text{Acetyl CoA} + CO_2 + H_2 \qquad (13)$$

It was during the study of this reaction that ferredoxin was discovered as an identifiable protein required as a reductant to link the oxidation of pyruvate to the production of hydrogen gas. The actual steps in the reaction are given in equations (14-16).

$$\text{Pyruvate} \xrightleftharpoons{\text{Thiamin pyrophosphate}} CO_2 + C_2 - \text{"acetaldehyde"} \qquad (14)$$

$$C_2 - \text{"acetaldehyde"} + Fd_{ox} + \text{CoA} \rightleftharpoons Fd_{red} + \text{Acetyl CoA} \qquad (15)$$

$$Fd_{red} \rightleftharpoons H_2 + Fd_{ox} \qquad (16)$$

The first part of the reaction has its counterpart in higher animal cells and involves the enzyme pyruvate dehydrogenase to form the 'active aldehyde' derivative from the vitamin-derived cofactor, thiamin pyrophosphate. In the anaerobic system the pyruvate dehydrogenase complex then reacts the active aldehyde with coenzyme A using ferredoxin as the electron acceptor (equation (15)). Ferredoxin is then reoxidized by a hydrogenase to produce H_2. Like many other electron-transferring proteins, the cytochromes for example, the ferredoxins are highly specialized enzymes in that they function as part of a total complex involving two or more other enzymes to carry out the complete reaction. While this review has concentrated on simple crystalline proteins for obvious chemical and physical reasons (our present methods are much better applied to them), in many cases the real biological system carrying out the complete reaction is beyond our capabilities at present in terms of obtaining the kind of detailed chemical information I have been covering and will discuss for the isolated component, ferredoxin, below. We should not lose sight of the fact that our final understanding will rest on knowledge of how the large integrated system works.

When it was found that ferredoxin supplied electrons of sufficient energy to reduce pyruvate, its function in the oxidation of other highly reducing compounds in anaerobic metabolism was examined and it was found to participate in a variety of such reactions including the production of H_2 from hypoxanthine, α-oxoglutarate, formate and acetaldehyde. These also require specific dehydrogenases in addition to ferredoxin.

The reduced ferredoxin can then be used to initiate a great variety of other reactions via the reduction of pyridine nucleotides or the reduction of hydroxylamine to ammonia. It can be involved in CO_2 fixation via the reductive carboxylation of acetyl CoA to pyruvate, it can reduce sulfite to sulfide, and in the presence of ATP can be used for the reduction of N_2 to

NH_3; the latter is the nitrogenase system (Table I) in which the iron-sulfur protein part of the system is somewhat more complex than the ferredoxins whose structures are given here as examples.

Photosynthetic Bacteria

The ferredoxin-dependent reactions occurring in these organisms, such as *Chromatium*, have certain analogies to the reactions discussed above for the anaerobic bacteria except it is now believed that a photoreduction of ferredoxin is added to the system as schematically represented in equation (17).

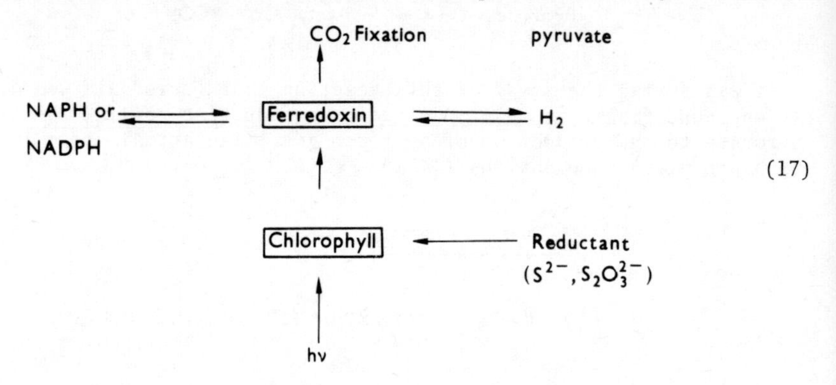

(17)

The precise function of several of the proteins in this class of enzymes has not been discovered. Although both rubredoxin and the high-potential iron protein from *Chromatium* are present in relatively high concentrations in the organisms, the exact reactions requiring reductants with the oxidation-reduction potentials of -0.057 V and +0.350 V (Table IV) are unclear as yet. Complete reviews should be consulted for details of the reactions in which the ferredoxins are involved.

Structure

A brief review of the chemical features of the iron in the ferredoxin-type proteins as discovered by physicochemical techniques applied either to solution or the frozen state is in order before presenting the findings from crystal structure. Spectroscopic investigation of these proteins has included the application of electron spin resonance[37,38], Mössbauer spectroscopy[39], optical spectroscopy (visible, IR, CD, ORD)[37], and NMR[40]. These approaches have now shown that strong antiferromagnetic exchange interactions exist between the component irons of the iron-sulfur clusters.

ESR Spectra

The non-heme iron of proteins of the ferredoxin class (containing inorganic sulfide with the exception of rubredoxin) was early discovered to be associated with an unusual ESR signal having the principal g value at $g = 1.94$ (Fig. 13). These signals appear in ferredoxin in the reduced state (for conditions, see refs. 10,37). Substitution of ^{57}Fe shows a broadening of these signals which establishes the ESR signal of the reduced protein as being associated with the Fe. In those proteins containing only two Fe atoms integration of the signal showed that the signal accounts for only half the Fe present. The signal intensity of one half that expected for two Fe atoms and the broadening observed with the substitution of ^{57}Fe is best accounted for if the unpaired spin density is shared by the two Fe nuclei, i.e. there is strong antiferromagnetic coupling between the Fe centers.

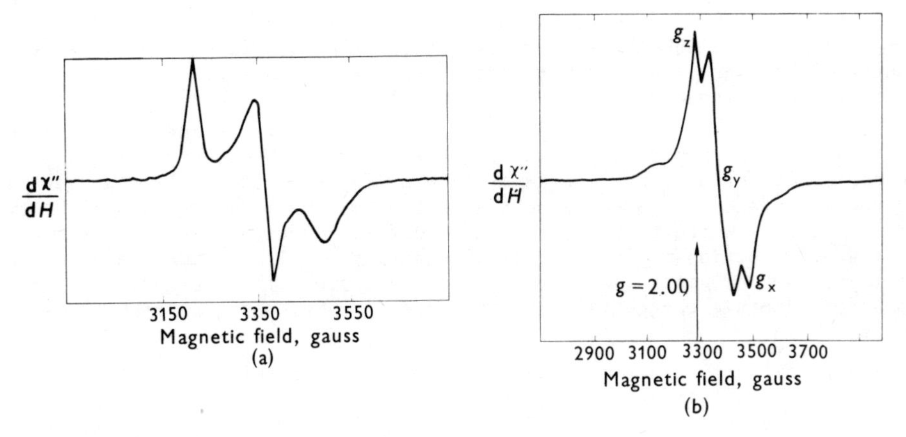

Fig. 13. ESR signals of reduced ferredoxins, spinach at left, *C. pasteurianum* at right.

Mössbauer Spectra

Mössbauer spectra have also been determined on the ferredoxins and are a subject of continuing investigation. Only the simplest case, a 2-Fe system (spinach ferredoxin) is shown here (Fig. 14).

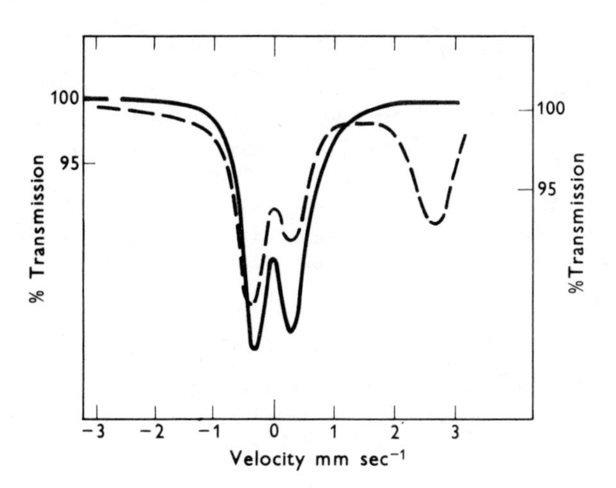

Fig. 14. Mössbauer spectra of (——) oxidised and (----) reduced ferredoxin.

The Mössbauer spectrum of the oxidized enzyme shows a narrowly split quadrupole pair characteristic of Fe(III). Only one type of Fe is present; thus all Fe atoms occupy similar environments. When the protein is reduced, a second pair of lines appears with much greater quadrupole splitting, characteristic of Fe(II). Since the new pair appears to account for only half of the Fe the Mössbauer spectrum is compatible with the interpretation that the oxidized protein contains two Fe(III) while the reduced protein contains the equivalent of one Fe(III) and one Fe(II). The exact interpretation of the oxidation state and spin-state of the iron, however, has been a matter of considerable discussion and other interpretations are possible[10].

For the ferredoxins containing more than two Fe atoms, the situation is more complicated. When the ferredoxin from *C. pasteurianum* (now known to contain 8 Fe) was reductively titrated, two ESR signals were observed which appeared to indicate the presence of two electron-transferring sites[41].

The NMR spectra of a number of ferredoxins have recently been examined and each class of Fe-S protein exhibits characteristic NMR spectra that depend on the redox state. The important feature of these spectra is that various proton resonances (apparently mostly from β-CH_2 and α-CH protons of the cysteine residues which bind to the Fe-S clusters) undergo contact interaction shifts and electron-nucleus dipolar broadening[42]. Temperature dependences of magnitudes of contact-shifted resonances reflect the presence of antiferromagnetic spin-exchange interactions between component Fe atoms of Fe-S proteins[42].

Lastly, primary structures of several of these proteins were available before the X-ray structure and the striking finding was the systematic occurrence of groups of cysteine residues as illustrated in Fig. 15 for the

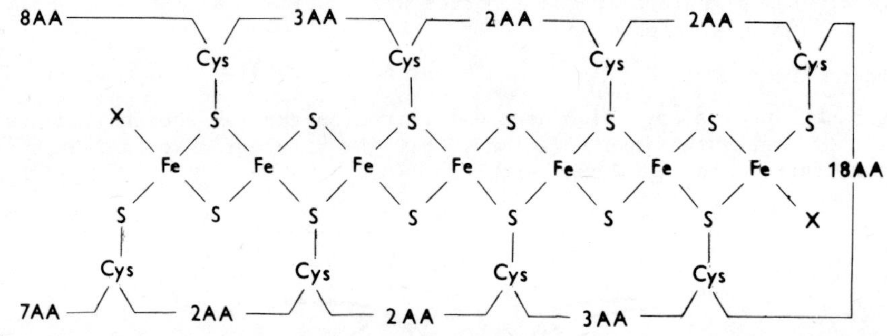

Fig. 15. Structure of the non-sulfur cluster in ferredoxin proposed on the basis of the amino acid sequence.

C. pasteurianum ferredoxin. This suggested the linear Fe cluster model shown especially since the analytical data indicated closer to 7 Fe atoms rather than 8. This is known to be incorrect; the Fe-S centers exist in two separate clusters.

Crystal Structure of Ferredoxins

There are now three crystal structures available on representatives of this class of enzymes which show increasing complexity; rubredoxin with 1 Fe, 4 Cys, and no S^{2-}; HiPIP with 4 Fe, 4 Cys and 4 S^{2-}; and ferredoxin from *M. aerogenes* with 8 Fe, 8 Cys, and 8 S^{2-}.

Rubredoxin[43]

This is a small protein of only 53 amino acids, and based, on the primary structure, the two-dimensional model in Fig. 16 was proposed[44]. The X-ray data have now been extended[45] to resolution of 2.0 Å and 1.7 Å. A refinement of this structure in the usual crystallographic sense (possible in small molecules) but not previously attempted with proteins, has now increased this resolution[45] to 1.5 Å. The general features of the molecule are indicated schematically in Fig. 17. The iron is approximately tetrahedrally coordinated to the S atoms of four cysteine residues. The exact bond lengths and angles are given schematically in Fig. 18 along with the experimental standard deviations. The most striking feature is that one of the Fe-S bonds standard deviations. The most striking feature is that one of the Fe-S bonds is considerably shorter than the others. The latter are near the bond lengths expected of Fe-S bonds in models. There is considerable distortion from strictly tetrahedral geometry.

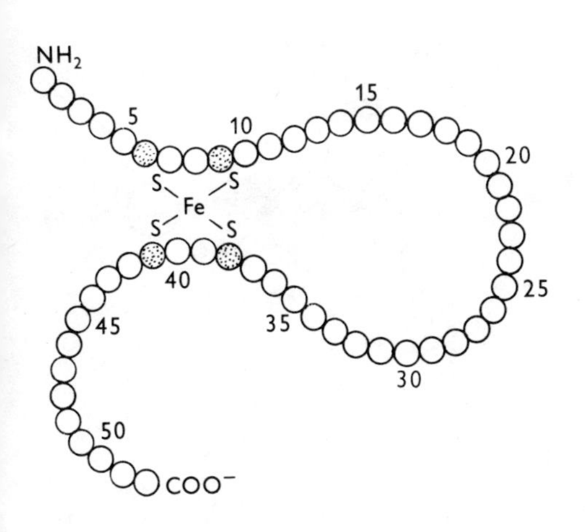

Fig. 16. Primary structure of rubredoxin.

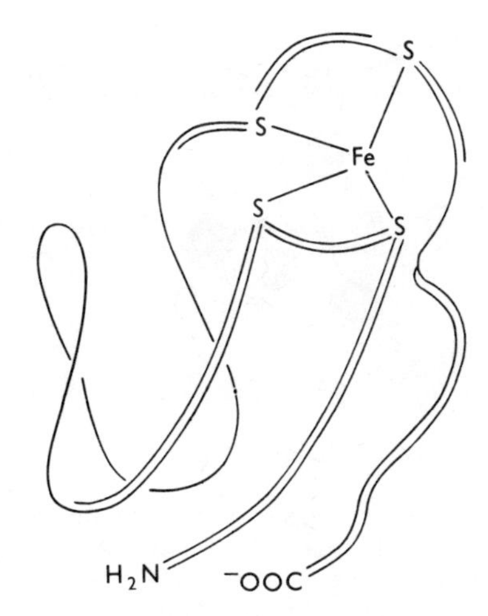

Fig. 17. Schematic of the three dimensional structure of rubredoxin.

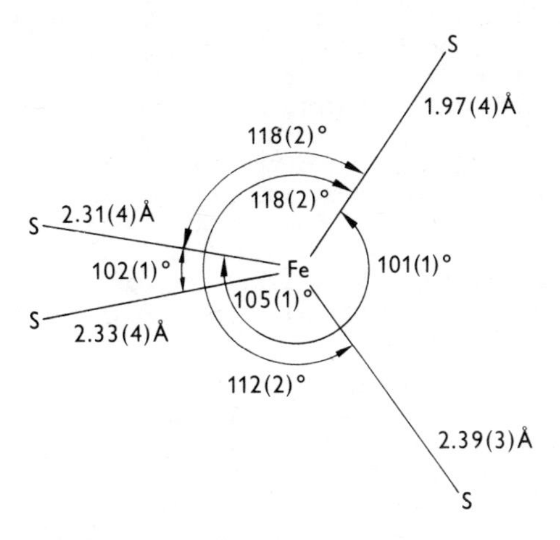

Fig. 18. Bond lengths and angles of iron-sulfur complex in rubredoxin. Standard deviations (in parentheses) are in units in the last place of the number to which each applies.

High-Potential Fe-S Cluster at 2.25 Å *Resolution*

This protein consists of an amino acid chain of 86 residues containing 4 cysteine residues. In addition it has 4 inorganic sulfide atoms included in the Fe-S cluster. The structure of the single Fe-S cluster in this protein is now available at 2.25 Å resolution and the results concerning the positions of the 4 Fe, 4 Cys, and 4 S^{2-} units are shown in Fig. 19[46]. This cluster is located near the center of an approximately spherical molecule, actually a prolate ellipsoid with axes of ≈35 Å and ≈20 Å. This cluster can best be described as two interlocking tetrahedra with a common center, the Fe atoms marking the vertices of one, the S^{2-} atom the vertices of the other. In other words an inorganic sulfide can be assigned to each face of the tetrahedron of 4 Fe atoms, each making 3 equivalent bonds to 3 Fe atoms.

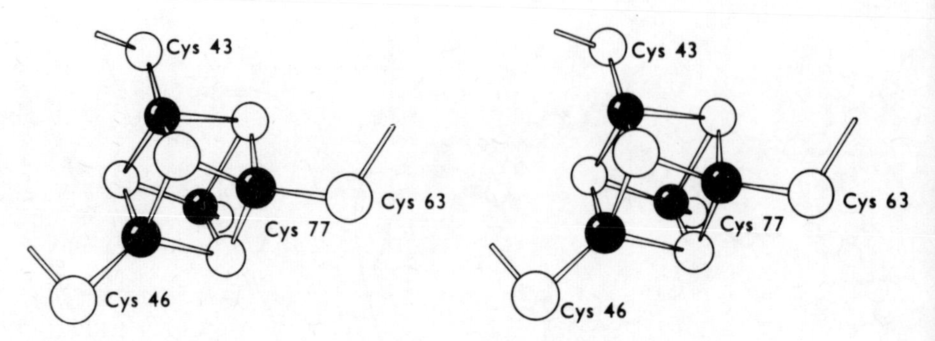

Fig. 19. Iron-sulfur clusters in *Chromatium* iron protein.

The cluster is located in what is apparently a highly hydrophobic region of the molecule. All but two of the apolar side-chains of the molecule line the interface where the two halves of the molecule come together to enclose the cluster. Unless there is some relaxation in the molecule in solution, van der Waals contacts would appear to prevent accessibility of solvent to the cluster.

Possible differences in structure between the oxidized and reduced forms of these proteins have not been examined in detail by X-ray crystallography as yet, but with this protein some preliminary conclusions are possible. The oxidation-reduction properties of this protein are such that crystals of the initially reduced protein undergo extensive oxidation during the collection of the X-ray data; thus the structure may represent an average from an unknown mixture of oxidized and reduced forms. However, data were collected on a fully oxidized form and there was exact isomorphism between these and the initially reduced crystals. Fourier difference maps calculated between the two forms also showed no significant differences at this resolutions. Thus there would appear to be very little recognition, as far as all but the most subtle features of structure are concerned, that the molecule has received or lost an electron.

The apparent equivalence of the Fe-atoms at this resolution needs comment. The mean Fe-S bond distance is 2.28 Å, typical of covalent Fe-S bonds. Fe-Fe distances are longer than those identified as Fe-Fe bonds in models (2.65 Å) and shorter than Fe-Fe distances identified as non-bonded 3.36 Å. At this resolution all the interatomic distances fall within two standard deviations of the mean value for each type; thus at this resolution there is no evidence for anything other than a tetrahedrally symmetric structure.

There are spectroscopic measurements that suggest there must be some deviation from exact tetrahedral symmetry. From the temperature dependence of the magnetic resonances assigned to the β-CH_2 protons of the cysteines, it can be concluded that in the oxidized state a paramagnetic hole remains localized on half the cluster for 10^{-4} sec, hence the 4 Fe atoms exist in two non-equivalent clusters[43,45]. Also detailed Mössbauer spectra suggest two slightly different Fe centers in the oxidized enzyme[46,47].

Crystal Structure of M. aerogenes Ferredoxin at 2.8 Å.

The *M. aerogenes* ferredoxin is similar to the enzyme from *C. pasteurianum* mentioned above. It contains 8 Fe, 8 cysteine residues, and 8 S atoms which appear as inorganic sulfide on denaturation. The primary structure of the protein is a peptide of 54 amino acids. A schematic of the structure deduced from the electron density map at 2.8 Å resolution is shown in Fig. 20[48]. The first point of general structure is that the Fe and S atoms are all located in two distinct clusters 12 Å apart. Thus the linear model mentioned

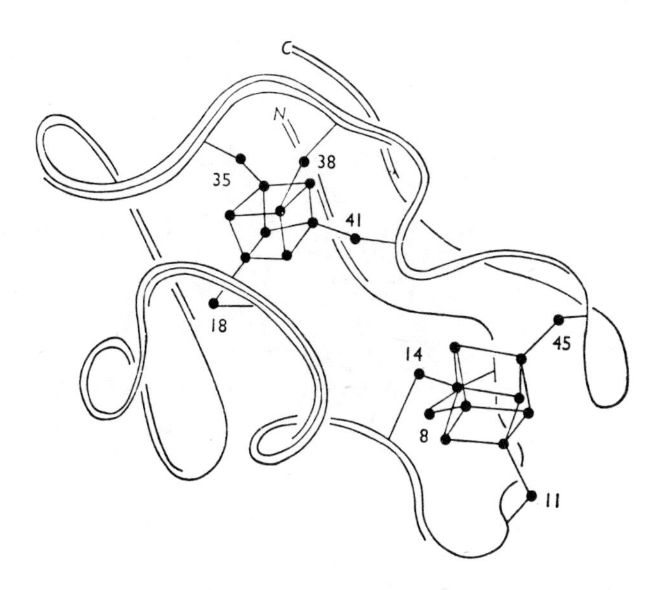

Fig. 20. Course of the main chain of *M. aerogenes* ferredoxin. The diagram is based on electron-density 'maps' at a resolution of 2.8 Å.

earlier cannot be true and the analytical data which were uncertain as to whether there were 7 or 8 Fe in this type of ferredoxin must reflect 8 Fe in each undenatured molecule. At this resolution the two Fe and S clusters each appear to have the same characteristics as the single Fe cluster just described for the high-potential Fe protein from *Chromatium*, although once again increased resolution may show significant changes from perfect tetrahedral symmetry for the two interlocking S and Fe tetrahedra. Residue 2 of this molecule is a tyrosine which has the same orientation to the first Fe-S cluster as Tyr 28 does to the second. This has suggested that these tyrosyls might play a part in the electron transfer[48].

It is somewhat harder to discuss mechanism in the case of these electron-transferring proteins, since one is asking what is the path from donor (which may be a substrate or another enzyme) to acceptor when we have no idea of the spatial relations among the three. It is of interest that the Fe-S clusters are apparently not directly accessible to the solvent. Thus there may be possible roles in electron transfer for groups on the protein such as the tyrosyls mentioned above. We may have a better chance of determining just what structural features of the protein are responsible for adjusting the oxidation-reduction potential of the Fe-S cluster, since detailed structures of proteins with both high and low oxidation potentials will be available at high resolution. It is possible that the distribution of the charged amino-acid side-chains in the vicinity of the clusters can have a marked effect on this potential.

Are there models in simple systems? There are in fact small molecular Fe-S complexes which provide close geometric analogies to the arrangement of the Fe and S atoms found in these ferredoxin clusters. Such clusters exist in the so-called cubane complexes such as $(C_5H_5FeS)_4$. Recently another related complex has been prepared, $(Et_4N)_2[Fe_4S_4(SCH_2Ph)_4]$, whose properties are very similar to those of the Fe clusters in the ferredoxins. The X-ray structure of this complex with the bond distances is shown in Fig. 21[40]. The Fe_4S_4 core is very similar to the structure of the clusters in the proteins just presented. At this high resolution there are obvious distortions from cubic

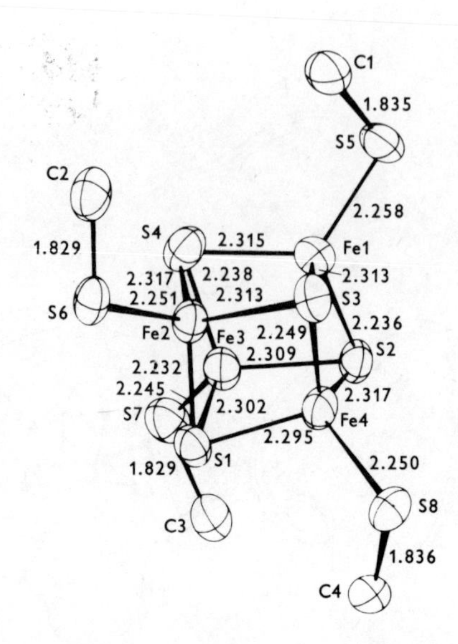

Fig. 21. The inner portion of the $[Fe_4S_4(SCH_2Ph)_4]^{2-}$ anion; 50% probability ellipsoids of thermal vibration are shown; hydrogen atoms are omitted for the sake of clarity.

symmetry in the model. The S-Fe-S angles are 104.1°, while the Fe-S-Fe angle are 73.8°, so that each face of the polyhedron is a rhomb. These rhombs are distinctly nonplanar. There is apparently no distinction between the Fe atom If the core S is considered as S^{2-}, the complex formally contains one pair each of Fe(II) and Fe(III); however, detailed studies suggest that a fully delocalized electronic description of the Fe_4S_4 core is a more meaningful description that the formal oxidation states 2 Fe(II), 2 Fe(III) and 4 S^{2-}.

As mentioned above, the ferredoxin-type proteins exist in two oxidation states differing by 1 e^- per 4 Fe cluster. Magnetic properties are consistent with a spin-singlet (S = 0) ground state for each such center in ferredoxin$_{(ox)}$ and high-potential iron protein$_{(red)}$. The model also undergoes a 1-electron oxidation-reduction compared to the proteins below: (equations 18-20).

$$\text{Ferredoxin}_{(red)} \rightleftharpoons \text{Ferredoxin}_{(ox)} + e^- \qquad E_{\frac{1}{2}} = -0.57\ \text{V} \tag{18}$$

$$\text{HiPIP}_{(ox)} + e^- \rightleftharpoons \text{HiPIP}_{(red)} \qquad E_{\frac{1}{2}} = +0.37\ \text{V} \tag{19}$$

$$[Fe_4S_4(SCH_2Ph)_4]^{3-} \rightleftharpoons [Fe_4S_4(SCH_2Ph)_4]^{2-} + e^- \qquad E_{\frac{1}{2}} = -1.19\ \text{V} \tag{20}$$

The ^{57}Fe Mössbauer spectrum of the oxidized model shows a single quadrupole-split doublet with splitting similar to that of the oxidized ferredoxins. There is no indication of two types of Fe, unlike the 8 Fe ferredoxins which appear to produce two closely overlapping pairs of doublets. The temperature dependence of the magnetic susceptibility of the model behaves much like the ferredoxins and shows the strong antiferromagnetic coupling of the Fe's.

The electronic spectra of the model and the proteins in the visible and near ultraviolet are also similar. These properties all suggest that the physico-chemical properties of the model closely approach those of the Fe-S cluster in the proteins. The precise value of the oxidation-reduction potential would be expected to be a function of the kind of environment in which the cluster is buried, a clear function of the protein.

The question of the presence or absence of conformational changes accompanying the addition or subtraction of the electron from proteins involved in oxidation-reduction reactions brings up some interesting points. When Fe is located in the highly restricted environment of the four nitrogen ligands of the heme ring system, the exact position of the Fe (in or out of the plane of the ring and bond distances to the *z*-ligands from the protein) is going to be very sensitive to the oxidation state and the spin state of the Fe (see Williams[49] for discussion). Thus it would not be too surprising if oxidation or reduction of the heme enzymes might be accompanied by significant conformational changes in the protein. The changes in hemoglobin accompanying oxygen binding and the change in Fe(II) from high-spin to low-spin have been well documented[50].

The X-ray structure of reduced cytochrome *c* is now available for comparison with the structure of the oxidized molecule, and there are some very significant conformational changes between the two oxidation states[51]. Parts of the main chain move to such an extent that almost a 2 Å decrease in the thickness of the molecule is produced. In the oxidized state the heme crevice appears to be relatively accessible to solvent, while in the reduced state the conformational changes are accompanied by a large rotation of a phenylalanine residue (Phe 82) such that it swings over and closes the heme cavity making it relatively hydrophobic and inaccessible to the solvent. How these rather large conformational changes on reduction of cytochrome *c* relate to function, the binding of the reducing enzyme, cytochrome oxidase, or the transfer of electrons between the three proteins is still a matter of speculation[51].

On the other hand, as mentioned above the conformational changes accompanying electron addition or withdrawal from the ferredoxin-type proteins may be minimal or nil. It has been pointed out by Williams[49] that such an irregular Fe-S cluster may accomodate the changes in bond angle and bond length required on oxidation state changes without any large change in the conformation of the surrounding protein, a prediction that would seem to be borne out at least by the preliminary data. The change could certainly be visualized as minimal if a single electron is shared by four Fe atoms without a formal oxidation state change being assignable to one of them.

This brief survey of catalysis by metalloenzymes has deliberately sacrificed some generality by discussing several specific examples in detail. This allowed a discussion at a molecular level which is not possible with many metalloenzymes at the present time. However, many of the approaches and general metal-protein interactions illustrated by these specific systems are likely to be found in other metalloenzymes, although detailed features will vary. For example, judging from the ubiquitous presence of the g = 1.94 ESR Fe signal, the Fe-S centers discussed above may be present in many enzymes involving a series of electron transfers. Zn-histidyl complexes may also be a rather general characteristic of Zn enzymes, at least those catalysing hydrolysis and hydration reactions. Zn-S interactions may possibly be present in others, e.g. alcohol dehydrogenase.

ACKNOWLEDGMENT

Original work carried out in the author's laboratory was supported by Grant AM 09070 from the National Institutes of Health and Grant GB-13344 from the National Science Foundation.

REFERENCES

Recent Reviews

COLEMAN, J.E. (1971). *Progress in Bioorganic Chemistry*, (eds. Kaiser, E.T. and Kezdy, F.J.), (Wiley, New York), pp. 159-344.

VALLEE, B.L. and WACKER, W.E.C. (1970). Metalloproteins in *The Proteins, Vol. 5*, 2nd edition, (ed. Neurath, H.), (Academic Press).

MILDVAN, A.S. (1970). *The Enzymes, Vol. 2*, (ed. Boyer, P.D.), (Academic Press, New York).

General References: Specific References to Systems Discussed

1. FLORKIN, M. and STOTZ, E.H. (eds.), (1965). *Comprehensive Biochemistry, Vol. 13*, (Elsevier, Amsterdam).

2. JENCKS, W.P. (1969). *Catalysis in Chemistry and Enzymology*, (McGraw-Hill, New York).

3. COTTON, F.A. and WILKINSON, G. (1966). *Advanced Inorganic Chemistry*, 2nd edition, (Interscience, New York), p. 761.

4. PHILLIPS, C.S.G. and WILLIAMS, R.J.P. (1965). *Inorganic Chemistry*, (Oxford University Press), p. 379.

5. BENKOVIC, S.J. and DUNIKOSKI, L.K. (1971). *J. Amer. Chem. Soc.*, 93, 1526.

6. BRESLOW, R. and OVERMAN, L.E. (1970). *J. Amer. Chem. Soc.*, 92, 1075.

7. VALLEE, B.L. and RIORDAN, J.F. (1968). *Brookhaven Symposia in Biology*, No. 13, p. 91.

8. HARTSUCK, J.A. and LIPSCOMB, W.N. (1971). *The Enzymes, Vol. 3*, 3rd edition, p. 1.

9. BEHNKE, W.D. and VALLEE, B.L. (1972). *Proc. Nat. Acad. Sci. U.S.A.*, 69, 2442.

10. COLEMAN, J.E. (1971). *Prog. Bioorganic Chem.*, 1, 159.

11. BRADSHAW, R.A. ERICSSON, L.H., WALSH, K.A. and NEURATH, H. (1969). *Proc. Nat. Acad. Sci. U.S.A.*, 63, 1389.

12. COLEMAN, J.E. and VALLEE, B.L. (1962). *J. Biol. Chem.*, 237, 3430.

13. VALLEE, B.L., RIORDAN, J.F., BETHUNE, J.L., COOMBS, T.L., AULD, D.S. and SOKOLOVSKY, M. (1968). *Biochemistry*, 7, 3547.

14. SIMPSON, R.T., RIORDAN, J.F. and VALLEE, B.L. (1963). *Biochemsitry*, 2, 616.

15. ROHOLT, O.A. and PRESSMAN, D. (1967). *Proc. Nat. Acad. Sci. U.S.A.*, 58, 280.

16. KOSHLAND, D.E. (1958). *Proc. Nat. Acad. Sci. U.S.A.*, 44, 98.

17. VALLEE, B.L., RIORDAN, J.F., JOHANSEN, J.T. and LIVINGSTON, D.M. (1971). *Cold Spring Harbor Symposia on Quantitative Biology*, 36, 517.

18. COLEMAN, J.E. and VALLEE, B.L. (1960). *J. Biol. Chem.*, 235, 390.

19. VALLEE, B.L., RIORDAN, J.F. and COLEMAN, J.E. (1963). *Proc. Nat. Acad. Sci. U.S.A.*, 49, 109.

20. KEILIN, D. and MANN, T. (1940). *Biochem. J.*, 34, 1163.

21. MAREN, T.H. (1967). *Physiol. Rev.*, 47, 595.

22. LINDSKOG, S., HENDERSON, L.E., KANNAN, K.K., LILJAS, A., NYMAN, P.O. and STRANDBERG, B. (1971). *The Enzymes, Vol. 5*, 3rd edition, p. 587.

22a. EDSALL, J.T. (1968). *Harvey Lectures* 62, p. 191.

23. KERNOHAN, J.C. (1964). *Biochim. Biophys. Acta*, 81, 346.

24. TASHIAN, R.E., PLATO, C.C. and SHOWS, T.B. (1963). *Science*, 140, 53.
25. POCKER, Y. and STONE, J.T. (1965). *J. Amer. Chem. Soc.*, 87, 5497.
26. KAISER, J.T. and LO, K-W. (1969). *J. Amer. Chem. Soc.*, 91, 4912.
27. POCKER, Y. and MEANY, J.E. (1965). *Biochemsitry*, 4, 2535.
28. LINDSKOG, S. and NYMAN, P.O. (1964). *Biochim. Biophys. Acta*, 85, 462.
29. REID, T.W. and WILSON, I.B. (1971). *The Enzymes, Vol. 4*, 3rd edition, p. 373.
30. KNOX, J. and WYCKOFF, H.W. (1973). *J. Mol. Biol.*, 74, 533; HANSON, A.W., APPLEBURY, M.L., COLEMAN, J.E. and WYCKOFF, H.W. (1970). *J. Biol. Chem.*, 245, 4975.
31. APPLEBURY, M.L., JOHNSON, B.P., and COLEMAN, J.E. (1970). *J. Biol. Chem.*, 245, 4968.
32. PERUTZ, M.F., MUIRHEAD, H., COX, J.M. and GOAMAN, L.C.G. (1968). *Nature*, 219, 131; see also PERUTZ, M.F. (1970). *Nature*, 228, 726.
33. TAYLOR, J.S. and COLEMAN, J.E. (1972). *Proc. Nat. Acad. Sci. U.S.A.*, 69, 859.
34. BUCHANNAN, B.B. (1966). *Struct. Bonding*, 1, 109.
35. BUCHANNAN, B.B. and ARNON, D.I. (1970). *Adv. Enzymol.*, 33, 119.
36. SAN PIETRO, A. (ed.), (1965). *Non-heme Iron Proteins*, (Antioch Press, Yellow Springs, Ohio).
37. PALMER, G., BRINTZINGER, H., ESTABROOK, R.W. and SANDS, R.H. (1967). In *Magnetic Resonance in Biological Systems*, (eds. Ehrenberg, A., Malmström, B.G. and Vänngård, T.), (Pergamon Press, Oxford), p. 159.
38. TSIBRIS, J.C.M., TSAI, R.L., GUNSALUS, I.C., ORME-JOHNSON, W.H., HANSEN, R.E. and BEINERT, H. (1968). *Proc. Nat. Acad. Sci. U.S.A.*, 59, 959.
39. BEARDEN, A.J. and MOSS, T.H. (1967). In *Magnetic Resonance in Biological Systems*, (eds. Ehrenberg, A., Malmström, B.G. and Vänngård, T.), (Pergamon Press, Oxford), p. 391
40. PHILLIPS, W.D. (1973). *Biochem. Soc. Trans., Abs. Comm.*, 1, 5; HERSKOVITZ, T., AVERILL, B.A., HOLM, R.H., IBERS, J.A. PHILLIPS, W.D. and WEIHER, J.F. (1972). *Proc. Nat. Acad. Sci. U.S.A.*, 69, 2437.
41. ORME-JOHNSON, W.H. and BEINERT, H. (1969). *Biochem. Biophys. Res. Commun.*, 36, 337.
42. PHILLIPS, W.D., POE, M., McDONALD, C.C. and BARTSCH, R.G. (1970). *Proc. Nat. Acad. Sci. U.S.A.*, 67, 682.
43. HERRIOT, J.R., SIEKER, L.C., JENSEN, L.H. and LOVENBERG, W. (1970). *J. Mol. Biol.*, 50, 391.
44. BACHMAYER, H., PIETTE, L.H., YASUNOBU, K.T. and WHITELEY, H.R. (1967). *Proc. Nat. Acad. Sci. U.S.A.*, 57, 122.
45. WATENPAUGH, K.D., SIEKER, L.C., HERRIOT, J.R. and JENSEN, L.H. (1971). *Cold Spring Harbor Symposia*, 36, 359.
46. CARTER, C.W., Jr., FREER, S.R. XUONG, Ng.H., ALDEN, R.A. and KRAUT, J. (1971). *Cold Spring Harbor Symposia*, 36, 381.
47 EVANS, M.C.W., HALL, D.O. and JOHNSON, C.E. (1970). *Biochem. J.*, 119, 289.
48. JENSEN, L.H. (1973). *Biochem. Soc. Trans., Abs. Comm.*, 1, 1.
49. WILLIAMS, R.J.P. (1971). *Cold Spring Harbor Symposia*, 36, 53.
50. PERUTZ, M.F. and TENEYCK, L.F. (1971). *Cold Spring Harbor Symposia*, 36, 295.
51. TAKANO, T., SWANSON, R., KALLAI, O.B. and DICKERSON, R.E. (1971). *Cold Spring Harbor Symposia*, 36, 397.

METAL ION-ACTIVATED ENZYMES

S. J. Benkovic
Department of Chemistry, Pennsylvania State University, University Park, Pennsylvania 16802, *USA*

In addition to the metalloenzymes with tightly bound metal retained during purification there are the related metal-activated enzymes[1]. The difference, however, is merely a quantitative one; the latter exhibit a lower affinity for the metal, so that the metal ion in general is lost during purification[2].

There is now considerable evidence which indicates that metal-activated enzymes and metalloenzymes catalyse reactions via ternary complexes consisting of a 1:1:1 ratio of enzyme:metal:substrate[3]. For the metal-activated enzymes four coodination schemes are possible:

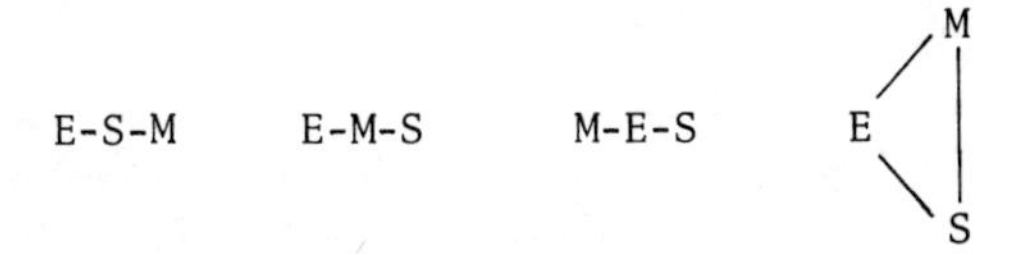

For metalloenzymes the substrate bridge complex is obviously not formed.

A variety of techniques is available for measuring the affinity and stoichiometry of enzymes, metals, and substrates including conventional dialysis and chromatographic methods; spectroscopic measurements utilizing fluoresecence, ESR, and paramagnetic enhancement of relaxation rates of magnetic nuclei, and X-ray crystallography[3]. From these methods, the type of ternary or quaternary complexes has been established for a number of enzymes. Furthermore such information permits inferences concerning the role of the metal ion in the catalytic process, although such projections are somewhat speculative since the above methods measure the characteristics of the more stable complexes, and not necessarily the catalytic one.

A number of cases of E-S-M ternary complexes have been established, particularly with phosphotransferase enzymes. Specifically for most kinases, e.g. adenylate, arginine and creatine kinase[4,5a,b] the reaction proceeds through a quaternary complex M-ATP-E-S_2 where M is either Mn^{2+}, or Mg^{2+}, S_2 the phosphoryl acceptor and ATP, adenosine triphosphate. The end products are phosphorylated S_2, and a dissociable E-ADP-M complex. For fructose diphosphatase which catalyses the hydrolysis of fructose-1, 6-diphosphate to fructose-6-phosphate, the quaternary complex may be represented[6] by M-ADP E-H_2O.

It is of interest to focus on the possible role of the metal ion in this catalytic process. This may be to activate the γ-phosphorus atom by σ- and π-electron withdrawal assuming that the Mn^{2+}-ATP complex in the enzyme is identical with that in aqueous solution, i.e. (writing Ad for adenosine):

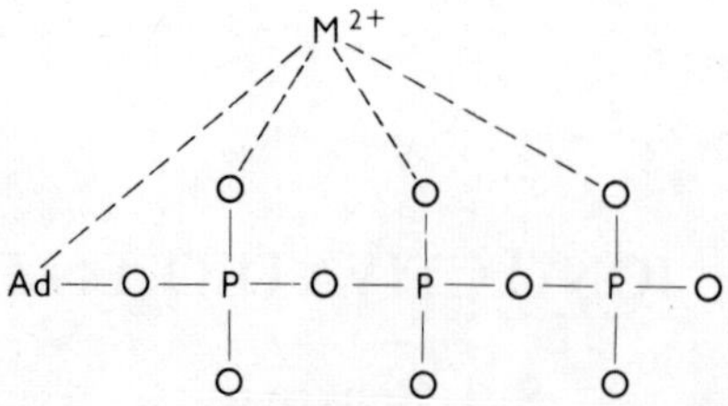

However, if all the decrease in free energy obtained by complexation (ca. $K \approx 10^3$) were reflected in the transition state for phosphoryl transfer, it would still be insufficient to account for the rate of the enzymatic reactions[7]. Moreover, since the kinase reactions are reversible to some extent, the dissymmetric location of the metal ion permits catalysis by the ion in only one direction if ground-state stabilities alone are considered, as shown below for the creatine kinase reaction proceeding in a metaphosphate mechanism:

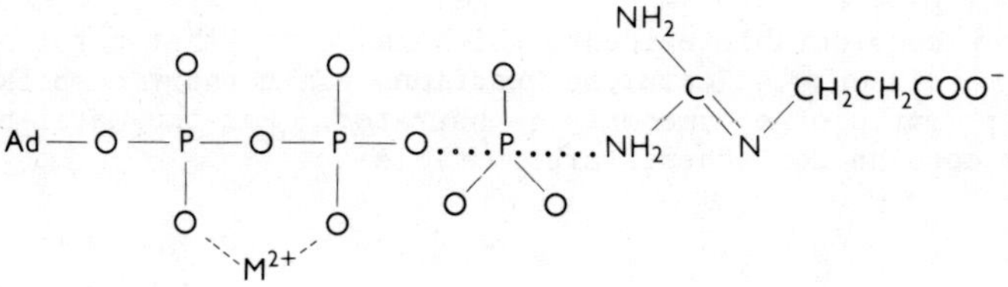

Of course catalysis of phosphoryl group expulsion from creatine phosphate may be promoted by acidic groups other than the metal ion on the enzyme so that symmetry of catalysis may be retained. Coordination of the metal ion to the incipient leaving group to catalyse phosphoryl group hydrolysis may also be important in the hydrolysis of fructose-1,6-diphosphate by fructose diphosphate, particularly as a result of recent NMR measurements of ^{13}C and ^{32}P relaxation rates in the presence of Mn(II) and enzyme which suggest Mn(II) interaction with the C-1 oxygen[6].

A second mode of catalysis would merely be the template effect, that is, proper orientation of the nucleophilic acceptor site and γ-phosphoryl moiety within the quaternary complex. A third unexcluded speculation is that the presence of the metal ion may permit the transfer to proceed via a pentacovalent intermediate, a species not generally observed in bimolecular displacement reactions on acyclic phosphorus metaphosphate (see below)[7].

Examples of metal-bridge complexes (E-M-S and $E<^{M}_{S}$) include both metalloenzymes and metal-enzyme complexes. In this category are carboxypeptidase[8], phosphoenolpyruvate carboxykinase[9], histidine deaminase[10], and muscle pyruvate kinase[11] to cite but a few. The formation of the metal-bridge complexes for many of these enzymes appears to occur through an S_N1 outer sphere mechanism[3]. For the metal-activated enzymes, any of the three possible formation pathways may operate, depending on the concentrations and rate constants for association/dissociation of the three species. Let us now consider several examples of catalytic mechanisms involving ternary metal-bridge complexes.

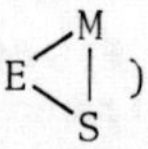

The active site of pyruvate kinase which catalyses the transfer of a phosphoryl moiety from phosphoenolpyruvate to ADP consists of two subsites: the E-Mn which binds the phosphoryl group and a second group which binds the carboxylate of pyruvate.

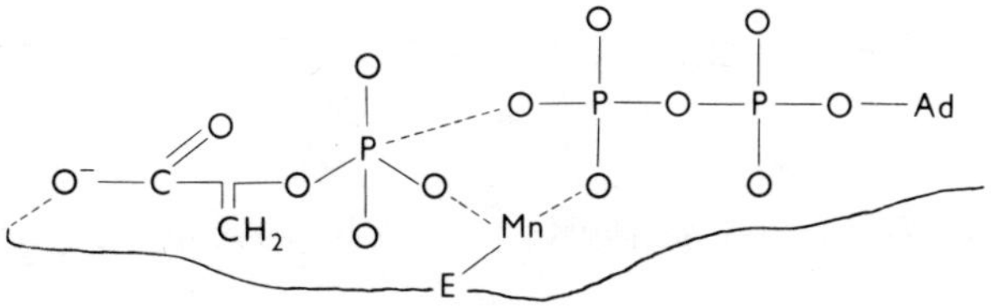

A possible role for the metal ion in catalysis is suggested by recent NMR measurements of the relaxation rates of various nuclei in an abortive complex whose dimensions are given below†:

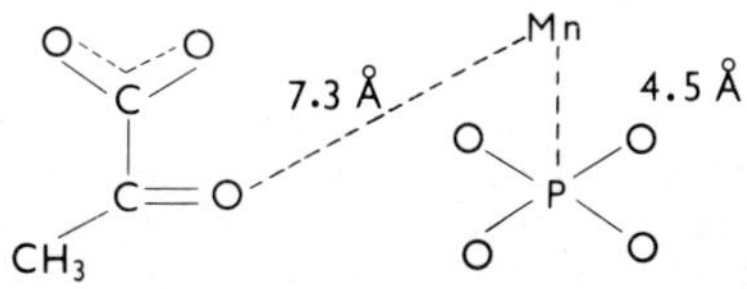

The distance between Mn and P is greater than that found for inner sphere Mn-O-P complexes where the distance is 2-3 Å. However, the Mn locus may accomodate formation of pentacovalent phosphorus species to be stabilized by coordination with the metal. Model studies indicate the feasibility of this mode of catalysis[7].

A more definitive example of a metal ion acting as a general acid catalyst — activating through σ-electron withdrawal — is found at the active site of histidine deaminase which catalyses the elimination of ammonia from histidine to form urocanate. The proposed mechanism may be depicted as:

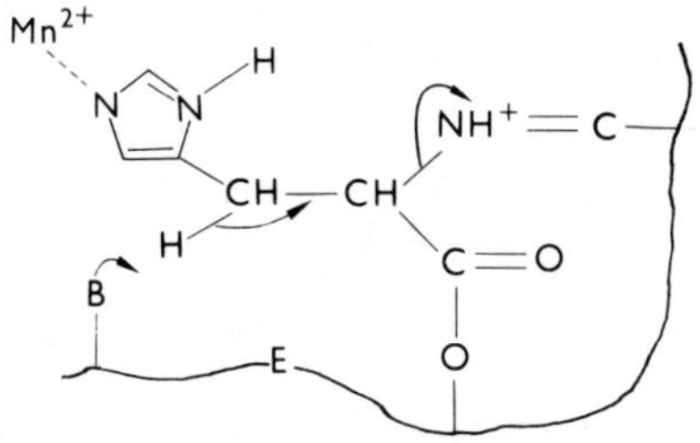

This means of activation by the metal ion to effect more facile formation of an incipient carbanion may apply to a number of enzyme-catalysed elimination reactions, e.g. enolase and D-xylose isomerase. The mechanism proposed for oxalacetate decarboxylase may likewise be viewed as a case where the metal ion may act as 'sink' for developing electron density in the transition state[12].

† Personal communication from A.S. Mildvan.

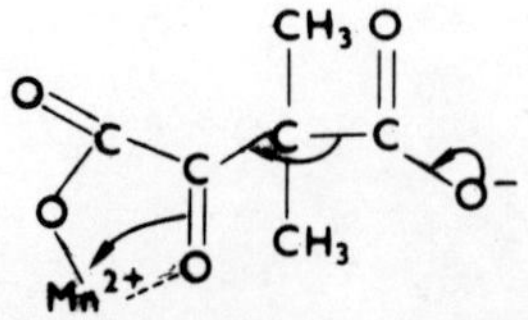

It should be noted that in the case of the E-M-S complexes of carboxypeptidase and carbonic anhydrase, the metal ion may function catalytically through a metal-hydroxide species[8,9,13], an alternative to mechanisms discussed above. These reactions, which are hydrolysis of an amide bond and hydration of carbon dioxide respectively, would occur at accelerated rates in the presence of the more nucleophilic hydroxyl group rather than a coordinated water molecule. Direct support for this suggestion, however, is stil lacking.

A final consideration is the role of metals in M-E-S complexes. Tightly bound metal ions in these systems may function primarily in a purely structura role, permitting the enzyme to assume a conformation associated with maximal activity. In the case of allosteric proteins these metal ions may serve as binding sites for modifiers of the enzymatic activity, as found for alkaline fructose-1,6-diphosphatase[6].

REFERENCES

1. VALLEE, B.L. (1955). *Adv. Protein Chem.*, 10, 317.
2. MALMSTRÖM, B.G. and ROSENBERG, A. (1959). *Adv. Enzymol.*, 21, 131.
3. MILDVAN, A.S. (1970). *The Enzymes, Vol. 2*, (ed. Boyer, P.D.), (Academic Press, New York).
4. COHN, M. and LEIGH, J.S., Jr. (1962). *Nature*, 193, 1037.

5a. O'SULLIVAN, W.J. and NODA, L. (1968). *J. Biol. Chem.*, 243, 1424.

5b. COHN, M. and HUGHES, T.R. (1960). *J. Biol. Chem.*, 235, 3250; (1962). 237, 176.

6. BENKOVIC, S.J., VILLAFRANCA, J.J. and LIBBY, C. (Unpublished results).
7. BENKOVIC, S.J. (1972). *Comprehensive Chemical Kinetics, Vol. 10*, (eds. Bamford, C.H. and Tipper, C.F.H.), (Elsevier, London).
8. LIPSCOMB, W.N., HARTSUCK, J.A., QUIOCHO, F.A. and REEKE, G.N., Jr. (1969). *Proc. Nat. Acad. Sci, U.S.A.*, 64, 28.
9. MILLER, R.S., MILDVAN, A.S., CHANG, H., EASTERDAY, R., MARUYAMA, H. and LANE, M.D. (1968). *J. Biol. Chem.*, 243, 6030.
10. GIVOT, I., MILDVAN, A.S. and ABELES, R.H. (1970). *Federation Proc.*, 29, 531.
11. MILDVAN, A.S., LEIGH, J.S., Jr. and COHN, M. (1967). *Biochemistry*, 6, 1805.
12. KOSICKI, G.W. and WESTHEIMER, F.H. (1968). *Biochemistry*, 7, 4303.
13. RIEPE, M.C. and WANG, J.H. (1968). *J. Biol. Chem.*, 243, 2779.

REDOX REACTIONS CATALYSED BY METALLOENZYMES

Gordon A. Hamilton
Department of Chemistry, Pennsylvania State University, University Park, Pennsylvania 16802, USA

Although not all enzymatic redox reactions require metal ion participation, every type of biological oxidation and reduction contains some examples which involve metalloenzyme catalysis. The four basic types of redox reactions encountered in biological systems[1,2], as well as specific examples of metalloenzyme catalysis of each type, are the following:

(1) Valence change of inorganic moieties or of heteroatoms

$$4Fe(II)Cyt.c + O_2 + 4H^+ \xrightarrow[Fe + Cu]{\text{cytochrome oxidase}} 4Fe(III)Cyt.c + 2H_2O$$

$$2O_2^- + H^+ \xrightarrow[\text{Cu + Zn or Mn}]{\text{superoxide dismutase}} O_2 + H_2O_2$$

$$SO_3^{2-} + H_2O + A \xrightarrow[Mo \pm Fe]{\text{sulfite oxidase}} SO_4^{2-} + AH_2$$

(2) Transhydrogenation

$$RCH_2OH + NAD^+ \xrightarrow[Zn]{\text{alcohol hydrogenase}} RCHO + NADH + H^+$$

$$RCH_2OH + O_2 \xrightarrow[Cu]{\text{galactose oxidase}} RCHO + H_2O_2$$

$$H_2O_2 \xrightarrow[Fe]{\text{catalase}} O_2 + H_2O$$

(3) Hydroxylation

$$R_3CH + O_2 + 2 \text{ reduced ferredoxins} \xrightarrow[\text{Fe}]{\text{cytochrome P-450}} R_3COH + 2 \text{ oxidized ferredoxins}$$

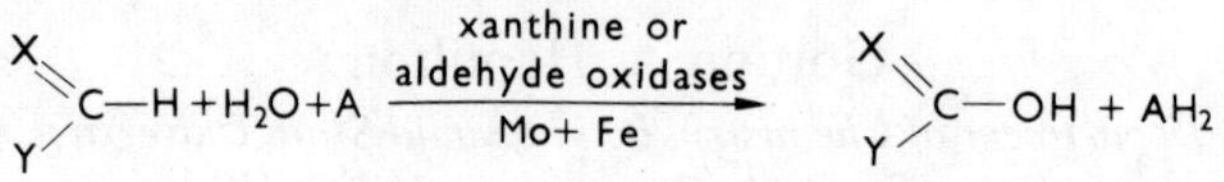

$$\begin{matrix}X \\ Y\end{matrix}\!\!>\!C{-}H + H_2O + A \xrightarrow[\text{Mo+ Fe}]{\text{xanthine or aldehyde oxidases}} \begin{matrix}X \\ Y\end{matrix}\!\!>\!C{-}OH + AH_2$$

(4) Oxidative cleavage of C-C bonds

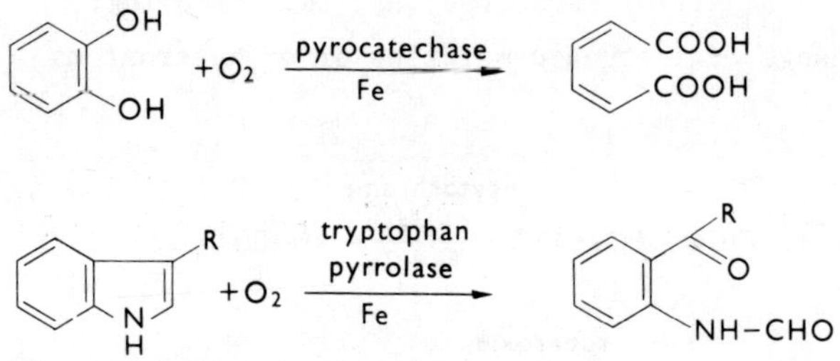

$$\text{catechol} + O_2 \xrightarrow[\text{Fe}]{\text{pyrocatechase}} \text{cis,cis-muconic acid (HOOC-CH=CH-CH=CH-COOH)}$$

$$\text{3-R-indole} + O_2 \xrightarrow[\text{Fe}]{\text{tryptophan pyrrolase}} \text{2-(NH-CHO)-C}_6\text{H}_4\text{-C(=O)R}$$

Some enzymatic redox reactions must be included in two or more of the above categories. A couple of metalloenzyme examples are the reactions catalysed by tyrosinase and *p*-hydroxyphenylpyruvate oxygenase illustrated below. Thus, although all enzymatic redox reactions can be included in the above categories, it is not always possible to include each individual reaction in only one category.

Tyrosinase (Cu)

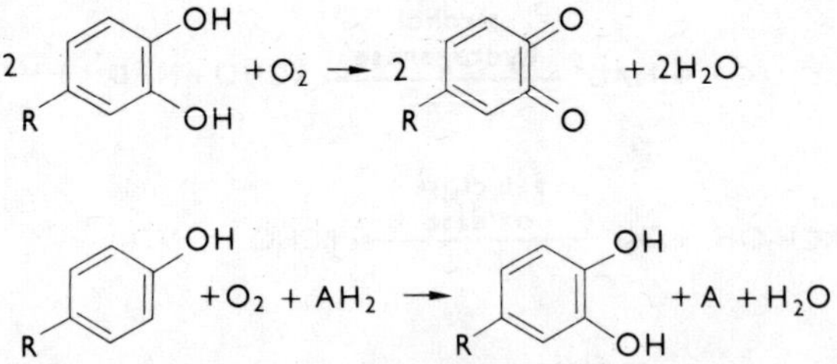

$$2\ \text{(4-R-catechol)} + O_2 \rightarrow 2\ \text{(4-R-o-quinone)} + 2H_2O$$

$$\text{(4-R-phenol)} + O_2 + AH_2 \rightarrow \text{(4-R-catechol)} + A + H_2O$$

p-Hydroxyphenylpyruvate oxygenase (Fe)

$$HO-C_6H_4-CH_2-CO-COOH + O_2 \rightarrow HO-C_6H_3(OH)-CH_2COOH + CO_2$$

Some metalloproteins have been given the title 'honorary enzymes'. They do not catalyse a reaction but, because they seem closely related to various enzymes, and are among the most thoroughly studied metalloproteins, they should be mentioned. Included in this group are (1) the electron-carrier proteins, such as the Fe-containing ferredoxins and cytochromes and the Cu-containing azurins, and (2) the oxygen-carrier proteins myoglobin (Fe), hemoglobin (Fe), and hemocyanin (Cu).

No X-ray structure of a metalloenzyme which catalyses a redox reaction has been reported. Thus, the detailed environment of the metal ion in most redox enzymes is largely unknown. However, it is clear that in many such enzymes the metal ion is complexed only to functional groups of the various amino-acids which make up the protein. In other redox enzymes the metal ion is in a more unique environment. Inorganic sulfide is a ligand of Fe not only in the ferredoxins[3,4] but also in the enzymes nitrogenase and xanthine oxidase as well as others. The carrier ring system[5] is unique to the cobalt-containing vitamin B_{12} coenzyme which is involved in catalysing various intramolecular redox and rearrangement reactions. The porphyrin ring system is present in many different types of metalloproteins including: (1) the oxygen-carrier proteins, myoglobin and hemoglobin; (2) the electron-carrier proteins such as cytochrome *c* and b_5; (3) the transhydrogenases, catalase and peroxidase; (4) the hydroxylase, cytochrome *P*-450; and (5) the oxidative-cleavage enzyme, tryptophan pyrrolase.

These ligands are probably intimately involved in catalysis by these enzymes, For example, the porphyrin ring system probably allows electron transfer to occur by a mechanism which does not require direct complexing of an external ligand to the metal ion; the X-ray structures[4] of the cytochromes indicate that the metal is buried in the molecule while the edge of the porphyrin ring system is available to the solvent and for interacting with other proteins. The extensive overlap and delocalization of the electrons of the Fe and the π system of the porphyrin would allow such a possibility. Also, the intermediates involved in catalase and peroxidase reactions are presumably stabilized by such delocalization. This has been considered in detail elsewhere[1].

The metal ion apparently performs several functions in the various redox reactions. Some of these appear to be the following: (1) to bind substrates to the enzyme, (2) to activate the enzyme by binding to remote sites, (3) to act as an acid catalyst to polarize bonds, (4) to transfer electrons with a change in valence of the metal ion, (5) to transfer electrons from one ligand to another with or without any change in metal-ion valence, (6) to complex with O_2 and allow it to react by an ionic mechanism, (7) to convert simple oxidants (H_2O_2, O_2, etc.) to more reactive metal-complexed species; and so on. In most enzymatic reactions combinations of the above are involved, and there are probably several additional but presently unknown roles performed by the metal ion. The first three roles listed above are not unique to redox enzymes; they are involved also in other metalloenzymatic reactions.

However, the last four functions are found only in redox enzymes. A detailed discussion of these has been given[1,2]. Given the added requirement that to be a catalyst the metal ion must be able to exchange its ligands rapidly it is thus perhaps not too surprising that usually only one metal ion will function in a given enzymatic reaction, and that only a very small number of all the possible metal ions are found to be involved in enzymatic redox reactions.

REFERENCES

1. HAMILTON, G.A. (1969). *Adv. Enzymol.*, 32, 55.
2. HAMILTON, G.A. (1971). In *Progress in Bioorganic Chemistry, Vol. 1*, (eds. Kaiser, E.T. and Kezdy, F.S.), (Wiley, New York), p. 83.
3. COLEMAN, J.E. (1971). *Ibid.*, p. 159
4. COLEMAN, J.E. (1973). See article in this volume.
5. ABELES, R.H. (1971). *Adv. Chem. Ser.*, 100, 346.

PART II
HETEROGENEOUS CATALYSIS

GENERAL SURVEY

Robert L. Burwell Jr.
Ipatieff Catalytic Laboratory,
Department of Chemistry, Northwestern University,
Evanston, Illinois 60201, *USA*

This article gives a general introduction to catalysis. It is followed by separate detailed articles on heterogeneous catalytic oxidation, on active sites, and on selectivity and poisoning.

Consider some process that occurs at a slow or negligible rate. The addition of a foreign substance provides the possibility for the occurrence of new elementary reactions. These new elementary processes provide a catalytic pathway if, and only if

(1) there is a cycle of elementary processes which regenerates the added substance (or reaction product thereof),

(2) the rates of all the elementary processes concerned are adequately fast.

Heterogeneous catalysis is distinguished from other types of catalysis by the characteristic that at least some of the intermediates are surface species adsorbed on a solid.

Table I provides examples of heterogeneous catalytic processes and the catalysts used for these processes.

CATALYTIC REACTORS

The reactions of Table I are effected most commonly in one of two types of reactors.

In the *static* or *batch* reactor, a fixed quantity of reactants is added to a vessel containing a catalyst. One follows the amounts of reactants and products with time. For example, a mixture of ethylene + H_2 could be admitted to a bulb upon the interior of which an evaporated film of Pt had been deposited.

In the *flow* reactor, the ethylene + H_2 is passed through a bed of catalyst at a set space velocity and the effluent analysed. Fig. 7 involves a flow reactor. Space velocity is the ratio

$$\frac{\text{quantity of reactants per unit time}}{\text{quantity of catalyst}}$$

Although a flow reactor may operate at a state which does not change with time, the concentrations of reactants and products change as one passes through the bed.

TABLE I

Examples of Heterogeneous Catalytic Reactions

Catalyst	Reaction	Temperature* °C
	HYDROGENATION REACTIONS AND THE LIKE	
Pt, Pd, Rh, Ni metals as such or supported on SiO_2, Al_2O_3, C	$H_2 + C_2H_4 = C_2H_6$	-100
	$H_2 + C_2H_2 = C_2H_4$	-100
	$H_2 + D_2 = 2\ HD$	-190
	$C_3H_8 + D_2 = C_3H_7D$, $C_3H_6D_2$, etc. + HD	50
	cyclopropane + H_2 = propane	40
	$C_2H_6 + H_2 = 2\ CH_4$	200
Pt, supported	H_2 + benzyl acetate = toluene + acetic acid	25
Cr_2O_3 activated to generate Cr^{3+} *cus* (coordinatively unsaturated surface)	$H_2 + C_2H_4 = C_2H_6$	-100
	$D_2 + C_3H_8 = C_3H_7D + HD$	200
	$H_2 + D_2 = 2\ HD$	-190
Metallic Fe	$3\ H_2 + N_2 = 2\ NH_3$	400
Pt, Cr_2O_3	heptane = toluene + 4 H_2	450
Pd, Cr_2O_3	1-butene = 2-butene, accompanying hydrogenation	0, 50
Pt, Ni, Cu	acetone + H_2 = 2-propanol	25, 25, 100
ZnO-Cr_2O_3	$CO + 2\ H_2 = CH_3OH$	450
Ni	$CO + 3\ H_2 = CH_4 + H_2O$	200
Basic oxides	acetone + 2-butanol = 2-propanol + butanone	150
	FORMATION OF LINEAR POLYETHYLENE	
Cr^{3+}/SiO_2	polymerization of C_2H_4 to linear polyethylene	50
	OLEFIN DISMUTATION	
Mo compounds on a support	2 propylene = ethylene + butylene	25
	DEHYDRATION	
Al_2O_3	$C_2H_5OH = C_2H_4 + H_2O$	300
	OXIDATION	
Metals, oxides, etc.	$H_2O_2 = H_2O + \frac{1}{2}\ O_2$	25 and up
Pt, many transition-metal oxides	$2\ H_2 + O_2 = 2\ H_2O$	0, 100+
	$2\ CO + O_2 = 2\ CO_2$	50, 200+
	$CH_4 + 2\ O_2 = CO_2 + 2\ H_2O$	200, 350+
Pt, V_2O_5/SiO_2	$2\ SO_2 + O_2 = 2\ SO_3$	400
Metallic Ag	$C_2H_4 + \frac{1}{2}O_2$ = ethylene oxide	200
Bismuth molybdate	$CH_3CH{=}CH_2 + O_2 \rightarrow CH_2{=}CH{-}CHO$	400
	$CH_3CH{=}CH_2 + NH_3 + O_2 \rightarrow CH_2{=}CHCN$	450
V_2O_5/SiO_2	benzene + O_2 → phthalic acid (COOH, COOH)	350
Fe_3O_4	$H_2O + CO = H_2 + CO_2$	450
$CuCl_2$/pumice	$2\ HCl + \frac{1}{2}O_2 = Cl_2 + H_2O$	375

ACIDIC CATALYSTS

Silica-aluminas, amorphous and crystalline (zeolites), exhibit strongly acidic surface sites as shown by bonding of bases and by indicators. They catalyse a variety of reactions which are usually considered to involve intermediates of a carbonium ion-like nature. Examples are: catalytic cracking of long-chain alkanes (450 °C), skeletal isomerization of alkanes (100 °C+), alkylation of aromatic hydrocarbons by olefins (25 °C+), and polymerization of butylenes to branched-chain rather low-molecular-weight materials.

* The listed temperature is the lowest at which a reasonable rate is obtained in a flow reactor. It is obviously just a crude indicator of catalytic reactivity.

A flow reactor operated at very low conversions is called a differential reactor.

CHEMISORPTION

In general, at least one reactant must be *chemisorbed* on the surface of the catalyst. In a few exceptional cases, *physical adsorption* suffices — for example, the interconversion of ortho- and parahydrogen can occur as a consequence of the mere presence of a hydrogen molecule in the magnetic field of paramagnetic surface atoms. Physical adsorption results from such forces as ordinarily lead to the liquefaction of gases, whereas chemisorption involves the formation of chemical bonds between adsorbate and solid.

Table II presents rough rate data for the rates of chemisorption of a number of simpler gases at about 0 °C upon metal films at low pressures, usually less than 0.001 torr. In principle, the films were clean but some of the data depends upon old experiments in which the films may have been contaminated and the rates may therefore have been in error.

TABLE II

Rates of Chemisorption on Metal Films, from Hayward and Trapnell (1964)

	Very fast	Slow	None to 0 °C
H_2	Ti, Zr, Nb, Ta, Cr, Mo, W, Fe, Co, Ni, Rh, Pd, Pt, Ba	Mn, Ge	K, Cu, Ag, Au, Zn, Cd, Al, In, Pb, Sn
O_2	all except Au		Au
N_2	La, Ti, Zr, Nb, Ta, Mo, W	Fe, Ba	as for H_2, also Ni, Rh, Pd, Pt
CO	as for H_2, also La, Mn, Cu	Al	K, Zn, Cd, In, Pb, Sn
CO_2	as for H_2, less Rh, Pd, Pt	Al	Rh, Pd, Pt, Cu, Zn, Cd
CO_4	Ti, Ta, Cr, Mo, W, Rh	Fe, Co, Ni,	
C_2H_4	as for H_2, also Cu, Ag	Al	as for CO

Much chemisorption under the conditions of Table II is very fast; nearly half of the collisions of the gas molecules with the surface result in adsorption. This is true even of adsorptions in which the final product is dissociatively adsorbed, e.g. H_2 on Pt, N_2 on W. Beyond surface coverages (θ) of about ½, the sticking coefficient declines, often rather rapidly. Some sticking coefficients are very small even at 0 °C and $\theta = 0$, e.g. N_2 on Fe, alkanes on many metals. Such chemisorption usually becomes faster as the temperature is increased; it possesses an activation energy, and it is called activated adsorption.

Application of such data to heterogeneous catalysis is far from straightforward. Conditions are too different. However, it provides some guide and it suggests that we may find catalytic reactions in which an adsorption or desorption step is rate-limiting and others in which a surface reaction is rate-limiting.

Reactions between hydrocarbons and hydrogen proceed via surface organometallic intermediates, some of which are shown in Fig. 1. All in all, quite an organometallic zoo is involved since evidence is strong not only for most of the types shown, but for still others.

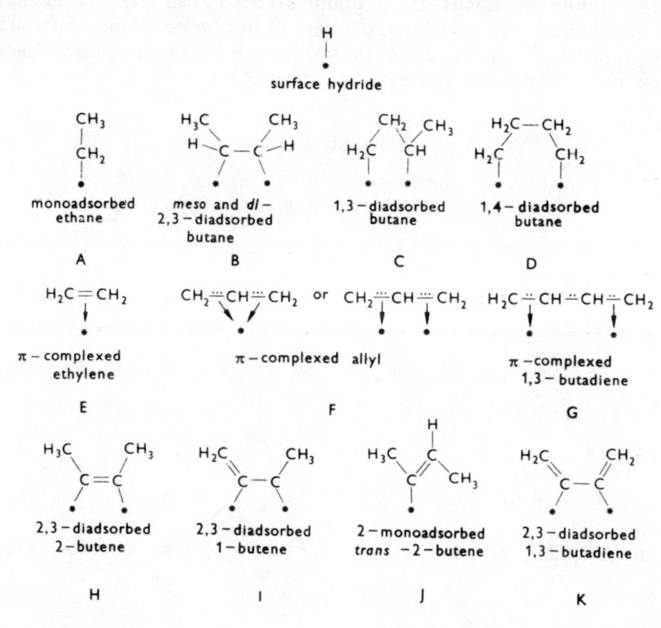

Fig. 1. Surface intermediates. This scheme shows some suggested intermediates in heterogeneous catalysis. The asterisk represents a surface site and adsorbed species are named by treating the surface site as a substituent.

The work of Sabatier is one of the major sources of modern heterogeneous catalysis. He was also the first to advance the general idea of surface organometallic intermediates. The organometallic aspect is not surprising since his unplanned entry into the area of heterogeneous catalysis occurred when he tried to make $Ni(CH_2{=}CH_2)_4$ in imitation of $Ni(C{=}O)_4$. However, the first specific proposals for surface organometallic intermediates, monoadsorbed alkane, 1,2-diadsorbed alkane, and 1-mono-adsorbed alkene (see Fig. 1, intermediates A,B and J), came in the 1930's and at a time when molecular organometallic compounds of transition elements were largely unknown and, consequently, forbidden. At least partly because the proposed surface organometallic intermediates were, therefore, somewhat suspect, little further activity in this area occurred until about 1950. Not long after, such surface species became respectable as inorganic chemists discovered how to make analogous molecular compounds of transition elements in quantity. A particularly good way to make such compounds is to stabilize the transition element in a low-oxidation number by the use of suitable ligands — exactly what the high-lattice energy of the metallic state does.

Classification of elementary reactions has proved useful in studying the mechanism of homogeneous reactions, both organic and inorganic. In organic chemistry one recognizes S_N1 and S_N2 displacement reactions, four-center reactions, radical combination reactions, and the like. A restricted classification of hypothetical elementary reactions in heterogeneous catalysis is shown in Fig. 2. The classification does not distinguish between breaking (or making) a σ-bond as in the dissociative adsorption of hydrogen (1la) or a π-bond as in the dissociative adsorption of ethylene to give $*CH_2CH_2*$ (IIc). Nor does the classification distinguish between a molecule in the gas phase and one which is physically adsorbed. Presumably, transfer between the two states is rapid and the kinetic consequences of the differences in state would be minor. A more detailed classification would consider this as well as other distinctions which are omitted in this simple classification.

In the Bonhoeffer-Farkas mechanism, the reaction $H_2(g) + D_2(g) = 2HD(g)$ proceeds by dissociative adsorption followed by random associative desorption.

$$D_2(g) + 2\bullet = 2D\bullet \text{ and } H_2(g) + 2\bullet = 2H\bullet \quad \text{(IIa)}$$

$$\text{followed by} \quad H\bullet + D\bullet = HD(g) + 2\bullet \quad \text{(-IIa)}$$

In the rival Rideal mechanism, associative desorption is assumed to be slow and the following reaction is assumed to lead to most of the exchange:

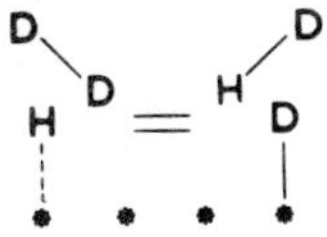

```
D            D
 \          /
  D   =   H
H           D
:           |
•    •   •  •
```
(Vc)

This reaction (Vc) is a gas adsorbate reaction. Despite a large amount of work on the subject, the mechanism of the hydrogen-deuterium exchange reaction is still in controversy particularly at temperatures of 100 to 200 °K.

Our knowledge of the exact nature of surface sites is slight. There is still controversy as to whether catalytic activity of metals is possessed uniformly by all surface atoms or whether it is associated with surface point defects, edges, dislocations, and the like.

Not surprisingly, then, our knowledge of the nature of the electronic binding in R-* is also weak. These problems are more closely tied to studies in chemisorption and their solutions are likely to come from studies in that field.

In the interim, fortunately, considerable mechanistic progress can be made without detailed knowledge about * and binding to it. Unfortunately, however, our ignorance of the nature of sites and the binding thereto makes it somewhat difficult to intercompare the three major manifestations of catalysis — homogeneous, heterogeneous, and enzyme. At one extreme, the surface chemistry of silica, alumina, and chromia resembles to a substantial degree the chemistry of model compounds which can be studied homogeneously. Useful intercomparisons are possible here. If, however, bonding should involve

Reaction	Bonds broken	Bonds Made
I Adsorption and its reverse, desorption		
a $* + NH_3(g) = H_3N*-$	none	$A-*$
b $* + H(g) = H*$	none	$A-*$
II Dissociative adsorption and its reverse, associative desorption		
a $2* + H_2(g) = 2H*$	A-A	$2(A-*)$
b $2* + CH_4(g) = CH_3* + H*$	A-B	$A-*, B-*$
c $2* + CH_2{=}CH_2(g) = *CH_2CH_2*$	A-A	$2(A-*)$
III Dissociative surface reaction and its reverse, associative surface reaction		
a $2* + C_2H_5* = H* + *CH_2CH_2*$	$A-B*$	$A-*, B-*$
b $2* + *CH_2CH_2CH_2CH_2* = 2*CH_2CH_2*$	$*A-A*$	$2(A-*)$
IV Reactive adsorption and its reverse, reactive desorption		
a $H* + C_2H_4(g) = C_2H_5*$	$A-*, B-B$	$A-B*, B-*$
b 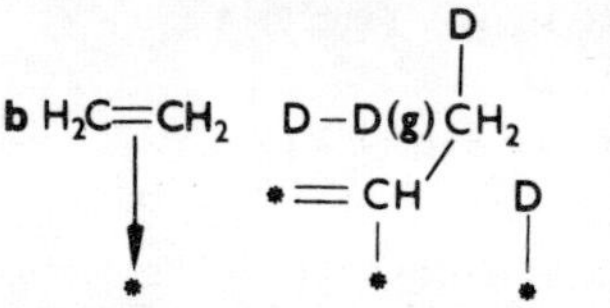 $H_2C{=}CH_2$ (→ $*$) $D-D(g)$ $*{=}CH(*)-CH_2D$, $D-*$	?, A-A	$A-B*, B-*, A-*$
c $*CH_2CH_2* + H_2(g) + * = *C_2H_5 + * + H*$	$B-*, A-A$	$A-B*, A-*$
V Gas-adsorbate reaction		
a $H(g) + H* = H_2(g) + *$	$A-*$	A-A
b $2H* + C_2H_4(g) = 2* + C_2H_6(g)$	$2(A-*), B-B$	2(A-B)
c $H* + D_2(g) + * = * + HD(g) + D*$	$A-*, B-B$ $A-*, B-C$	$B-*, A-B$ $E-*, A-B$
d $D* + H_2C{=}CH{-}CH_3(g) = DH_2C{-}CH{=}CH_3(g) + H*$	D-E	C-D

Fig. 2. A restricted classification of elementary reactions.

collective electrons — that is, electrons from orbitals which extend throughout substantial volumes of the solid — analogies with small molecular catalysts will be poor. Studies of metal cluster compounds could be useful here.

Any solid is coordinatively unsaturated in the sense that the creation of solid–vacuum interfaces requires the performance of work. A species constituting the solid will adsorb and become part of the solid with consequent reduction in free energy (when the sublimation pressure of the solid is negligible). Suitable molecules or atoms should also adsorb at the sites for crystal growth. This idea that sites for adsorption and heterogeneous catalysis involve coordinative surface unsaturation is an old idea and one which has been qualitatively useful. Indeed, much of the versatility of heterogeneous catalysis probably results from the fact that a surface of necessity usually involves coordinative unsaturation. However, this idea may not be directly applicable to all examples of heterogeneous catalysis, for example, to solid proton acid catalysts.

Table II summarizes chemisorption on 'clean' metals. To some degree, one may consider the surface atoms of metals to be coordinatively unsaturated. There has been considerable controversy as to the nature of binding at a metal surface. Is a surface hydrogen atom best considered to be bound somewhat as in transition metal hydrides like $HCo(CN)_5^{3-}$ (emphasizing surface coordinative unsaturation)? Or, does the bonding involve the collective electrons of the metal (not emphasizing surface coordinative unsaturation)? That is, is the bonding localized or not? Recent opinions have tended to favor the localized model but the matter is far from resolved.

The metallic bond is strong in catalytically useful metals such as nickel and platinum. Separation of a crystal of these metals into individual atoms is endothermic by 100 kcal. or more per gram-atom. One expects, therefore, strong chemisorption to the surfaces of these metals. But, for example, are monoadsorbed cyclopentane and hydrogen (formed by $C_5H_{10} + 2* = C_5H_9* + H*$) positioned over single surface atoms or are they between and bonded to several? We do not really know, although the structures of interstitial carbides and nitrides and the fact that the next metal atom chooses the second alternative suggest that adsorbed species are often bonded to two or more surface atoms. Dense packing of identical sites, which is likely to occur on metallic surfaces, favors multisite processes like IIc. On alumina and chromia, however, equivalent sites are probably farther apart and multisite processes should be less favored.

Binding at Cr^{3+}(*cus*) where (*cus*) stands for 'coordinatively unsaturated surface' can be considered as completing the coordination sphere of Cr^{3+} and to be analogous to binding in coordination complexes. The pair sites, Fig.16, are (*cus*) species, Cr^{3+}(*cus*) and O^{2-}(*cus*). The adsorption of hydrogen on this pair site involves heterolytic dissociative adsorption with a proton going to O^{2-} to generate surface OH^- and a hydride ion becoming bound at Cr^{3+}. The species CH_2DCH_2 bound to Cr^{3+} is formally $CH_2DCH_2^-$. Kokes has shown that hydrogenation of ethylene on ZnO proceeds by a very similar mechanism.

Pearson's concept of hard and soft acids and bases (SHAB) is qualitatively useful in considering adsorption.

Other things being equal, a soft acid prefers to combine with a soft base; a hard acid with a hard base. Soft and hard must be distinguished from weak and strong. A strong, soft base may combine well with a hard acid, e.g. S^{2-} with H^+.

In general, 'soft' indicates highly polarizable and large; hard, less polarizable and small. The prototype of the hard acid is the proton, of the hard base, the hydroxide ion.

Among bases, the order of increasing hardness is

$$F^- > Cl^- > Br^- > I^-,$$

$$O \gg S > Se > Te \quad \text{as in } H_2O,$$

$$N \gg P > As \quad \text{as in } NH_3.$$

Small, highly charged ions are hard acids; large, less charged ions are soft acids: Li^+, Mg^{2+}, Al^{3+}, Fe^{3+}, Cr^{3+}, are hard, also CO_2 and SO_3. Cu^+, Ag^+, Pt^{2+} are soft, also tetracyanoethylene, ethylene, CO, H^- and C_2H_5. The higher the charge on a metal ion, the harder it is. Thus Co^{3+} and Fe^{3+} are hard, Co^{2+}, and Fe^{2+} are intermediate. F^- and H_2O bond much more strongly to Al^{3+} than I^- or S^{2-}, but I^- bonds more strongly to Ag^+.

The surfaces of transition metals behave as M(0) and are primarily soft acids. R_2S, R_3P, CN^-, CO bond strongly; R_2O binds weakly.

Thus, water should bind strongly to Cr^{3+}(*cus*) and, indeed, water is a strong poison for chromia. Ethylene and CH_3^- should bind weakly. H^+ and CO_2 should bind strongly to O^{2-} but one would hardly expect ethylene to bind.

Poisoning is a pervasive problem in heterogeneous catalysis. The SHAB concept is useful in predicting the kinds of compounds which would be poisons of a particular catalyst.

DIFFUSION

On present views, a heterogeneous catalytic reaction involves diffusion of the reactants from the bulk fluid phase (gas or liquid) to the surface of the catalyst, adsorption, reactions of the type shown in Fig. 2, desorption, and finally diffusion of the products to the bulk phase. Many catalysts are supported catalysts. For example, Pt/SiO_2 involves often about 1% by weight of Pt on SiO_2. The silica is silica gel which consists of tiny spheroids of silica (diameter about 150 Å) cemented together by Si-O-Si bonds. About half of the volume of a particle of silica-gel is void space. The Pt occurs as small crystallites supported throughout the cemented silica-gel particles whose diameters vary from 0.05 to several mm depending upon the reaction and conditions under which the catalyst will be used. The diameter of the crystallites of Pt will usually be in the range 15 to 100 Å depending upon the conditions of preparation of the Pt/SiO_2.

If the reactant is to diffuse from the bulk fluid phase into the silica-gel particle, there must be a concentration gradient of the reactant within the porosity of the silica-gel such that the concentration of the reactant steadily decreases from the external surface of the particle into the interior. If this gradient becomes large, the concentration of the reactant in the void space in the center of the particle becomes significantly less than the concentration at the external surface and it may even fall to zero. In such cases, the catalytic effectiveness drops and the rate becomes significantly less than one expects on the assumption that the concentration throughout the particle is that in the external fluid phase.

In addition, marked concentration gradients can affect the selectivity of a catalytic reaction, that is, the proportions in which products appear. Fig. 3 shows one way in which this effect may appear. The concentrations of reactants and products are plotted vertically against the distance of penetration into the pores which are shown in an idealized fashion as cylindrical

In mechanistic studies, it is essential that the concentration gradients be small but this has not been true of many studies reported in the literature. The gradients will be assumed to be small and of negligible effect on rate and selectivity in what follows.

KINETICS OF HETEROGENEOUS CATALYTIC REACTIONS

If the sites for adsorption are identical and noninteracting in a simple adsorption,

$$A(g) + * = A*$$

and if θ is the fraction of sites $*$ converted to A and $(1 - \theta)$ is the fraction

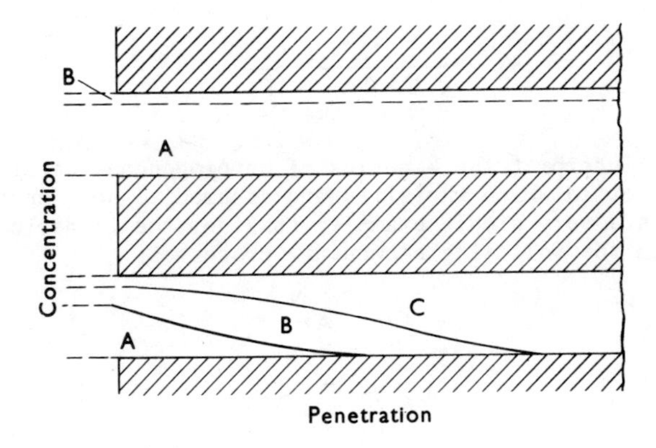

Fig. 3. The effect of concentration gradients upon selectivity. The reaction is A → B → C for which —C≡C— + $2H_2$ = —HC═CH— + H_2 = —H_2C—CH_2— would serve as an example. In this case, the concentration of hydrogen does not change much if it is in substantial excess. In the upper Figure, diffusion is fast relative to the rates of reaction and the effects of concentration gradients are small. In the lower Figure, diffusion is slow, concentration gradients are large. In the first case, the concentration of B is never large enough to generate significant amounts of C. In the second case, the far interior of the pores is entirely C, and C appears in the gas phase. One could incorrectly conclude the C was an initial product.

of free sites, then at equilibrium,

$$K = \frac{\theta}{(1 - \theta)p_A} \qquad \text{or} \qquad \theta = \frac{Kp_A}{(1 + Kp_A)} .$$

This is Langmuir's adsorption isotherm and θ *v.* p_A is shown graphically in Fig. 4. Note that the model implicitly assumes that the heat of adsorption of A is independent of θ.

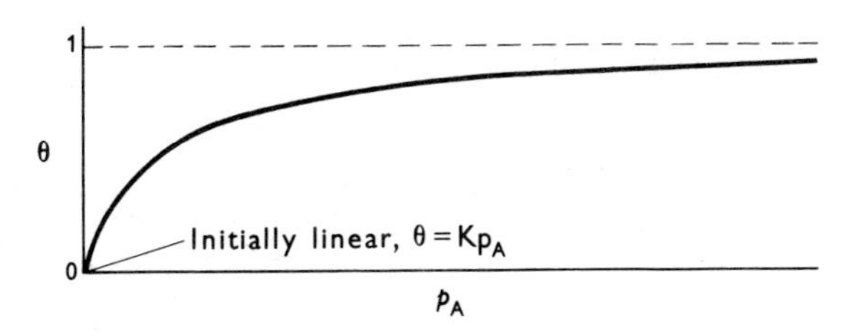

Fig. 4. Variation of surface coverage θ with the pressure of the adsorptive A according to the Langmuir Adsorption Isotherm.

A similar equation relates the concentration of substrate to the fractional coverage of sites on an enzyme.

When two gases adsorb competitively on the same sites, if θ_A is the fraction of sites covered by A and K_A is the equilibrium constant for adsorption of A,

$$\theta_A = \frac{K_A p_A}{1 + K_A p_A + K_B p_B} .$$

Most attempts to interpret the kinetics of heterogeneous catalytic reactions involve a model in which the sites are assumed to be identical and noninteracting. As a very simple example, consider the isomerization of A to B proceeding via the following elementary processes:

$$A + * = A* \quad (1)$$

$$A* = B* \quad (2)$$

$$B* = B + * \quad (3)$$

The kinetic representation of even such a simple system is rather complicated but let us look at two special cases at such low conversions that the reverse reaction can be ignored.

(a) Reaction (1) (adsorption of A) is slow and irreversible, reactions (2) and (3) are fast and at equilibrium, and θ_A and θ_B are very small. The rate of A = B is then the rate at which A adsorbs. This is proportional to p_A and to θ_*, the fraction of free sites. θ_* here is essentially equal to 1 since coverages by A and B are assumed very small. Then,

$$\text{rate} = k_1 p_A.$$

(b) Steps (1) and (3) are at equilibrium and reaction (2) is rate-limiting and irreversible. The rate is proportional to θ_A for which an expression was given above.

$$\text{rate} = \frac{k_2 K_A p_A}{1 + K_A p_A + K_B p_B} .$$

At low enough conversion, $K_B p_B$ can be ignored and

$$\text{rate} = \frac{k_2 K_A p_A}{1 + K_A p_A} ,$$

and if K_A is small enough,

$$\text{rate} = k_2 K_A p_A ,$$

which is of the same form as in (a) but the rate constant has a different interpretaion.

If K_A is large enough, θ_A becomes nearly equal to unity, $K_A p_A \gg 1$, and

$$\text{rate} = k_2.$$

When K_A or p_A is very small, the reaction is first-order in A, and when K_A or p_A is very large, zero-order.

If K_B is large, the reaction is inhibited by B much more than one would expect merely from the effect of the reverse reaction.

Such procedures can be extended to more complex mechanisms.

However, there are three problems. First, the sites on the surface of the catalyst may not be identical and, indeed, this should be a common situation. Conventional catalysts, say Pt/Al_2O_3, often involved very tiny crystallites of the active component in the range of 15 to 50 Å in diameter. By the nature of the preparation, there might well be a relatively high concentration of edges, steps or other sites of coordination lower than in the center of the more densely packed planes. Further, of course, more than one crystal plane is apt to be present. Such multiplicity of types of sites leads, in effect, to the presence of more than one catalytic species.

A priori, then, one might expect that the rate constant per unit area of active component would depend upon the details of the preparation of the catalyst. For example, one might expect the rate of hydrogenation of some olefin on Pt/SiO_2 to depend upon the average crystallite size of the Pt and the details of the preparation of the catalyst. If such variation is observed, the reaction is said to be *structure-sensitive* or *demanding*. If, however, the reaction rate per unit area is independent of crystallite size and details of preparation, the reaction is said to be *structure-insensitive* or *facile*. The investigation of this problem is one of considerable current interest and the work of Boudart may be particularly noted. The matter is not as yet unravelled. Some reactions such as the hydrogenation of ethylene on Pt appear to be structure-insensitive, but other reactions appear to be structure-sensitive. No general understanding has as yet emerged.

Secondly, the sites may be interacting. For example, the heat of adsorption of A may decrease with increasing θ_A even if the sites are uniform. Thus, K_A would not be constant but would decrease with increase of θ_A. Thirdly, in low-energy electron diffraction on clean metal surfaces (at very low pressures) behavior resembling that of the Langmuir adsorption isotherm is rarely seen. Often the behavior may be of some type like that shown in Fig. 5.

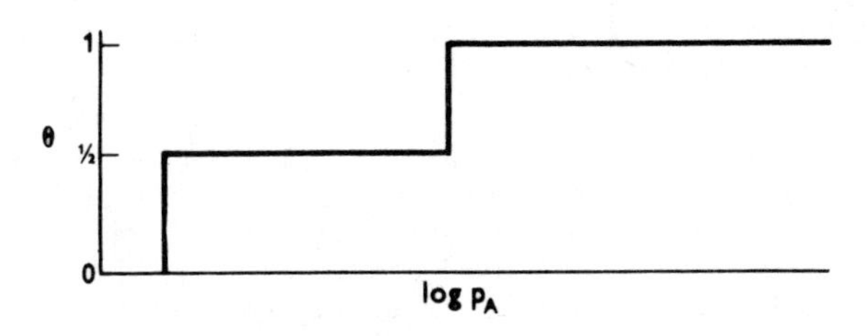

Fig. 5. The type of variation of coverage with p_A sometimes observed by low energy diffraction on single crystal surfaces.

Here, the transitions from $\theta_A = 0$ to $\theta_A = \frac{1}{2}$ and from $\theta_A = \frac{1}{2}$ to $\theta_A = 1$ resemble phase changes and involve discontinuous jumps in θ_A as p_A changes.

The degree to which the second item influences kinetics in specific cases is rarely completely understood but we do have considerable information relating to the problem. However, there has been virtually no progress in understanding the apparent conflict between the adsorption isotherms measured on conventional catalysts at pressures in the vicinity of one atmosphere and the isotherms observed in the low-energy electron diffraction study of one plane of a single crystal at pressures in the range of 10^{-4} to 10^{-10} torr.

MECHANISM

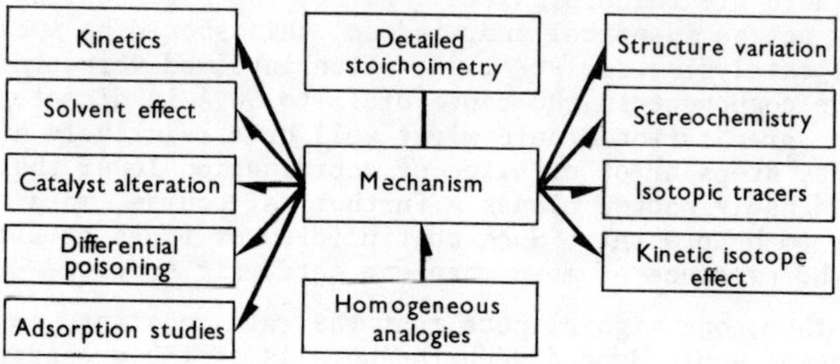

Kinetics
Solvent effect
Catalyst alteration
Differential poisoning
Adsorption studies
Detailed stoichoimetry
Mechanism
Homogeneous analogies
Structure variation
Stereochemistry
Isotopic tracers
Kinetic isotope effect

Fig. 6. Data used in assigning mechanism.

Fig. 6 indicates the kinds of data which have been used in the assignment of mechanism in heterogeneous catalysis, that is in the elimination of all possible mechanisms except one (in the rare ideal case). In preliminary work the detailed stoichiometry should be determined and there should be a search for any detectable gas-phase intermediates. Homogeneous analogies suggest mechanism, but experimental data test mechanism.

A mechanism which is verified by just one type of data is certainly 'not well established'. Kinetics is a key aspect of mechanism and kinetics is a key test of mechanism. In my opinion, however, *mere kinetics* has been a less powerful eliminator of mechanisms in heterogeneous catalysis than in homogeneous reactions. Too many different mechanisms give similar kinetics (possibilities are enlarged considerably over homogeneous reactions by the adsorption-desorption steps and by the possibility of reactive adsorption and desorption steps of Fig. 2), and fine distinctions among different kinetics require an exact reproducibility of rate measurement which is not easily achieved. Various attempts have been made to force fit all kinetics into a restricted group of possible mechanisms of the Langmuir-Hinshelwood type. One assumption of this model is that ΔH of adsorption reactions like IIb ($2* + CH_4(g) = CH_3* + H*$) is independent of the degree of coverage although this must be a poor approximation in many cases.

Consider the hydrogenation of olefins. *Mere study of the kinetics* of the hydrogenation of just ethylene has not provided enough information for any reliable establishment of the mechanism involved. Further information is essential. Kinetic studies of several compounds of appropriately different structures — a series of substituted ethylenes — would be helpful. Reactants which add stereochemical data are particularly powerful in eliminating mechanisms (Siegel). Competitive studies of two or more reactants present together may well prove useful because this kind of study largely frees one from the problem of exact catalyst reproducibility. The use of deuterium as well as hydrogen would give us the data classified under isotopic tracers and, perhaps, the kinetic isotope effect.

Hydrogenation of Nitrogen

Hydrogen is a component of most systems for which useful mechanisms can be proposed, a situation occasioned by the spur of the technical importance of a number of hydrogenation reactions and by the ready availability of deuterium as an isotopic tracer. The hydrogenation of nitrogen was one of the first reactions subjected to detailed mechanistic examination. The purist may scorn the ammonia synthesis reaction as too applied, but the real problem is that this reaction is applied on a large scale only in developed countries. The intimate relation of this reaction to the survival of what we can recognize as civilization may justify our examining its mechanism in some detail.

The basic problem in nitrogen fixation is the strength of the third bond of the nitrogen-nitrogen triple bond, not the weakness of the nitrogen-hydrogen bond, as shown in Table III.

TABLE III

Bond Energies, from Cottrell (1958)†

	Triple	Double	Single
N-N	226	100	39
N-H			93
C-C	194	143	83
C-H			99
H-H			104

Soil organisms can activate molecular nitrogen at ambient temperatures, but no present man-made catalyst can accomplish this. Os, Ru, and Fe catalyse the reaction

$$N_2 + 3H_2 = 2NH_3 \quad \Delta H_{25} = -21 \text{ kcal} \tag{1}$$

but temperatures of 400 °C or so are needed to get practical rates, and high pressures are necessary at these temperatures because equilibrium in the exothermic reaction (1) shifts to the left. Technically, one uses iron with the addition of *promoters*, in this case, alumina, and a basic oxide like potassium and calcium oxides. The alumina seems to function mainly by reducing sintering with concomitant loss of surface area; the basic oxide functions by reducing the sensitivity of the catalyst to poisons and probably by other means as yet not understood.

† *The Strengths of Chemical Bonds*, 2nd edition, (Butterworths, London).

Adsorption of hydrogen on the catalyst is fast and the hydrogen-deuterium exchange reaction occurs rapidly at room temperatures. Thus, activation of hydrogen does not seem to limit the rate. On the other hand, adsorption of nitrogen is slow and isotopic exchange between $^{14}N_2$ and $^{15}N_2$ occurs rapidly only above 400 °C. The approximate identity of rates of adsorption of nitrogen, of the isotopic exchange reaction, and of the ammonia synthesis reaction suggests strongly that the slow step is the activation of nitrogen. This relation can readily be understood in terms of the bond energies.

Desorption of ammonia is not a slow step since isotopic exchange between ammonia and deuterium on iron is much faster than the rate of ammonia decomposition. Thirty years ago, Temkin and Pyzhev advanced the following mechanism to represent these facts:

$$3/2H_2(g) + 3* = 3H* \quad (2)$$

$$1/2N_2(g) + * = N* \quad (3)$$

$$3H* + N* \overset{(4)}{=} 2H* + * + HN* \overset{(5)}{=} H* + 2* + H_2N* \overset{(6)}{=} NH_3(g) + 4*$$

The slow step is assumed to be (3). The detailed model is not of the Langmuir-Hinshelwood type. Further, the heat of adsorption of nitrogen is presumed to decrease linearly with surface coverage.

Recently, there has been a considerable revival of interest in the mechanism of ammonia synthesis. The Temkin and Pyzhev model is not universally accepted in all details although the general outline seems well established. As Ozaki, Taylor and Boudart have shown, under certain conditions the kinetic isotope effect observed in replacing hydrogen by deuterium is not compatible with the original Temkin and Pyzhev model. Some think that process (4) is also slow and that the following reactions may play a part:

$$N_2(g) + 2* = *N{=}N* \quad (7)$$

$$*N{=}N* + 2H* = 2HN* + 2* \quad (8)$$

Hydrogenation of Acetylenes and Dienes

At room temperatures most transition metals catalyse the hydrogenation of acetylene, but only to ethylene or ethane. Methane is formed only under much more drastic conditions. This contrast between the hydrogenations of HC≡CH and N≡N may be surprising at first glance. However, the σ (single) bond between carbon and carbon is strong and each additional π bond is increasingly weaker, as shown in the Table of bond energies. But in nitrogen the σ bond is weak and each π bond is progressively stronger. Thus the π bond of acetylene is easily opened, but the σ bond is not.

To eliminate problems arising from the cleavage of the hydrogen-carbon bond, let us consider the hydrogenation of dimethylacetylene (2-butyne) rather than that of acetylene itself, and let us use Pd as the catalyst because it gives particularly simple results. Fig. 7 shows what happens if one passes 2-butyne in excess hydrogen over a dilute dispersion of Pd on alumina. The reaction is very fast — 10 to 100 molecules of dimethylacetylene react per second per surface atom of Pd. The chemistry of the reaction on other forms of Pd is similar.

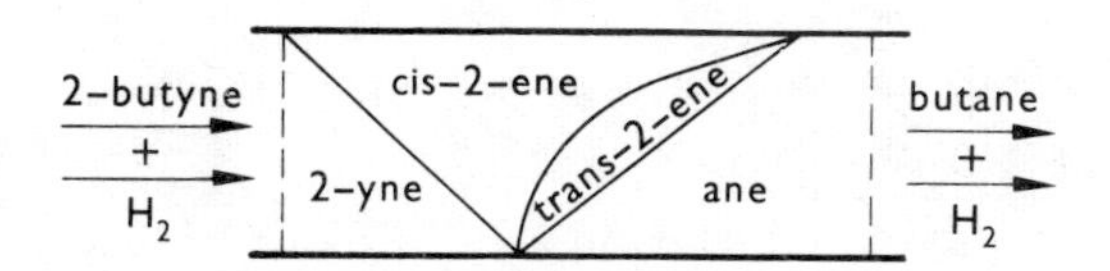

Fig. 7. Reaction between dimethylacetylene (2-butyne) and excess hydrogen on a Pd/Al_2O_3 catalyst. Once all acetylene is gone, the initial product, *cis*-2-butene, isomerizes and hydrogenates to butane. The catalyst bed is confined between the vertical dashed lines, and the relative amounts of reactants and products are plotted vertically.

Pd is unusually *selective* in that reactions (11), (12) and (13) are slow in the presence of 2-butyne.

$$CH_3C{\equiv}CCH_3 + H_2 = cis\text{-}CH_3CH{=}CHCH_3 \qquad (10)$$

$$CH_3C{\equiv}CCH_3 + 2H_2 = CH_3CH_2{-}CH_2CH_3 \qquad (11)$$

$$cis\text{-}CH_3CH{=}CHCH_3 = trans\text{-}CH_3CH{=}CHCH_3 \qquad (12)$$

$$CH_3CH{=}CHCH_3 + H_2 = CH_3CH_2{-}CH_2CH_3 \qquad (13)$$

The reaction is *stereoselective* in that only the *cis* isomer is formed in reaction (10). 2-Butyne is so strongly adsorbed that it excludes 2-butene from the surface, as one could predict from the Table of bond energies. Reactions (12) and (13) commence only after butyne, which is a poison for these reactions, has disappeared by reaction (10). Initial hydrogenation of 2-butyne, (10), is little poisoned by dimethyl sulfide, but further reaction of the products of reaction (10) (i.e. reactions (12) and (13)) is strongly poisoned by dimethyl sulfide. Thus, if one adds a little dimethyl sulfide to the mixture of 2-butyne and hydrogen, reactions (12) and (13) remain poisoned even after all 2-butyne has reacted as shown in Fig. 8. This gas-phase system resembles that much used for the preparation of *cis*-olefins by liquid-phase hydrogenation on supported Pd poisoned by lead acetate and quinoline (Lindlar's catalyst). There too, reactions (12) and (13) remain largely poisoned after the disappearance of the acetylene.

The hydrogenation of 2-butyne seems to proceed via the intermediates shown in Fig. 9. Evidence for this mechanism derives from stereochemistry and from isotopic tracer investigations in which the position of the deuterium atoms that were introduced was determined by a combination of mass spectroscopy and NMR.

If a mixture of H_2 and D_2 is passed over the catalyst at room temperatures, the isotopic exchange reaction,

$$H_2 + D_2 = 2HD$$

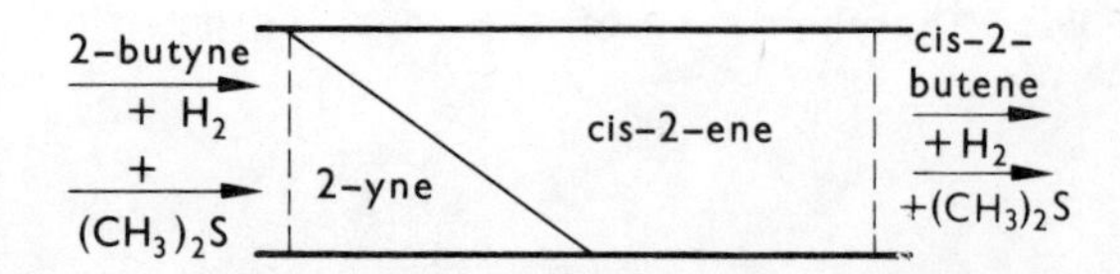

Fig. 8. Reaction between dimethylacetylene (2-butyne) and excess hydrogen on a Pd/Al_2O_3 catalyst in the presence of a small amount of dimethyl sulfide. The initial reaction of dimethylacetylene is slightly poisoned, but subsequent reaction of *cis*-2-butene remains strongly poisoned even after all acetylene has disappeared.

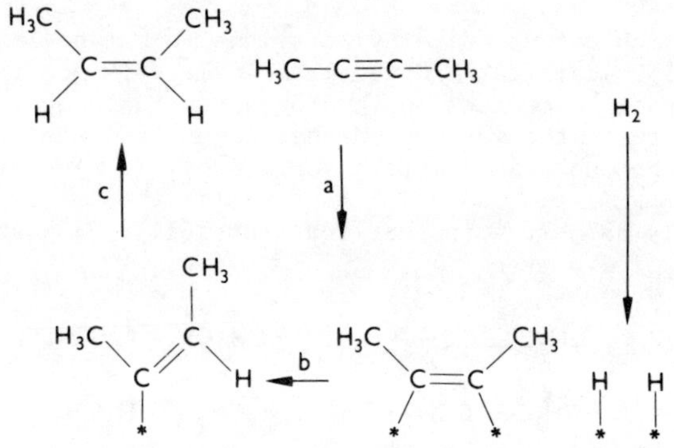

Fig. 9. Hydrogenation of 2-butyne.

goes nearly to equilibrium. The addition of 2-butyne almost completely suppresses formation of HD in the gas phase. The isotopic distribution of the *cis*-2-butene which is formed accords with random addition of hydrogen and deuterium atoms, not molecules. Mono- and diadsorbed butenes must largely cover the surface during the hydrogenation of butyne. Hydrogen adsorbs at sites at which crowding is too great to permit adsorption of the much more strongly bonded butyne. The rate of desorption of hydrogen is negligible. In the reaction between deuterium and methylallene more than 95% of the 2-butene formed is 2-butene-d_2. This result shows that reaction -b has a rate which is negligible with respect to that of reaction b.

The exact nature of the binding in these species is unsettled, particularly that of the product of reaction a. However, its geometry seems well established and to resemble the structure of fairly close molecular analogs, shown in Fig. 10.

Hydrogenations of butynes and butadienes appear to be more complicated on other metals of Group VIII. Subsequent reactions of *cis*-2-butene (reactions (12) and (13)) remain poisoned as long as butyne or butadiene remains unreacted but direct reaction of butyne to butane, reaction (11), occurs to a substantial degree and some *trans*-olefin is an initial product.

Hydrogenation of Olefins

There is wide but not universal agreement that hydrogenation of olefins on metallic catalysts proceeds basically via the Horiuti-Polanyi mechanism, which we can exemplify by the propylene-propane-deuterium system shown in Fig. 11. Dissociative adsorption of deuterium, $2* + D_2(g) = 2D*$ (IIa), is assumed to occur simultaneously.

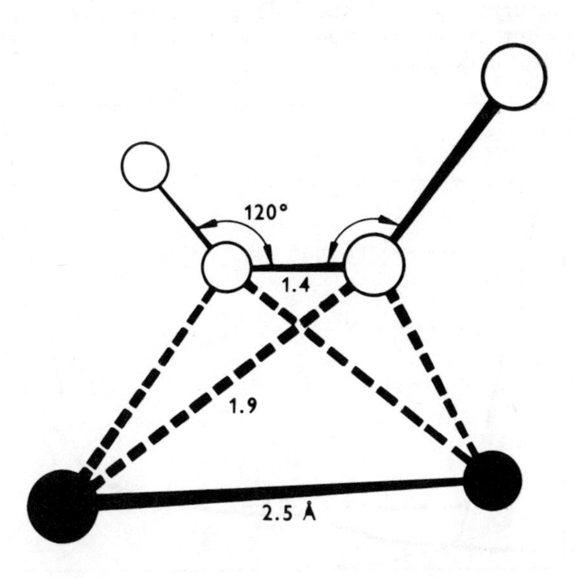

Fig. 10. The geometry of binuclear transition-metal complexes with acetylenes. The filled circles are cobalt, the open ones are carbon. Each cobalt is bonded to three carbonyl groups which are not shown. In Bailey, Churchill, Hunt, Mason and Wilkinson (1964), the two-bonded carbon atoms are those in perfluorocyclohex-1-yne-3-ene. In Sly (1959), the acetylene is diphenylacetylene. In a somewhat more complicated tetranuclear cobalt complex, Dahl and Smith (1962), the acetylene residue also has the geometry of a disubstituted ethylene.

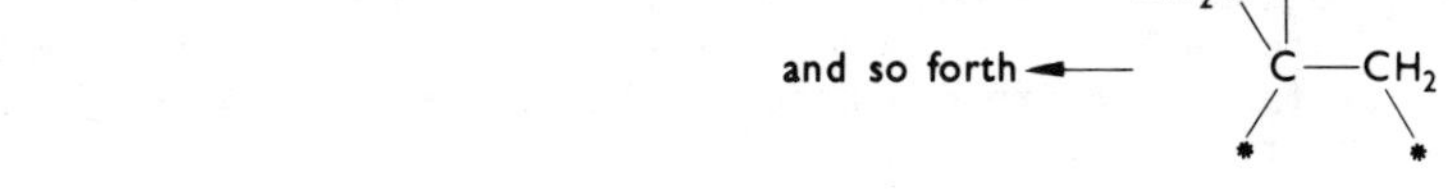

propane-d_0 propane-d_1 propylene propane-d_2
IIb IIc −IIIa −IIb k_3 k_{-2}
H_3C H CH_3 H_3C H C—CH_2 H_3C H CH_2D
DH_2C H C—CH_2
and so forth

Fig. 11. The Horiuti-Polanyi mechanism. See Fig. 2 for numbering.

In the original Horiuti-Polanyi mechanism, IIa + IIc, -IIIa, -IIb, would give propane-d_2. However, -IIIa and IIc are readily reversible and there are additional side reactions. These complications lead to smeared isotopic distributions in the alkane product and to isotopically exchanged and isomerized olefins. Thus the hydrogenation of olefins is more complicated in detail than that of acetylenes and dienes.

Dissociative adsorption of alkanes, IIb, is slow relative to the hydrogenation of olefins, but it occurs readily under conditions a little more drastic than those used for hydrogenation.

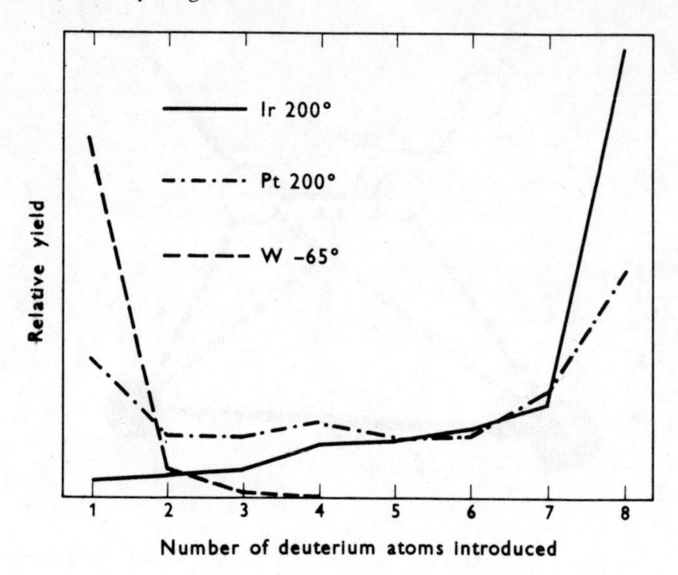

Fig. 12. The distribution patterns of the initial products of isotopic exchange between propane and deuterium on metallic catalysts. Rh at -24 °C resembles Pt at 200 °C. Pd at 100 to 200 °C is intermediate between the Pt and Ir curves. At -46 °C Ni resembles the W curve, but the degree of extensive multiple exchange increases at higher temperatures. Although the support has some effect upon the distribution patterns, evaporated metal films and supported catalysts of the same metal give rather similar patterns. In this diagram, k_{-2}/k_3 is small on W, but large on Ir. From Addy and Bond (1957) and Kemball (1954).

Fig. 12 shows the isotopic distributions of the initial products of isotopic exchange between propane and deuterium on several catalysts. On Pt, Rh, Ir, Pd, and, at higher temperatures, Ni, most of the first molecules which desorb contain many atoms of deuterium. The distributions observed at higher conversions are modified by isotopic dilution of the deuterium and by readsorption of previously exchanged alkanes.

As first suggested by Kemball, the multiply-exchanged alkane results from reversal of step -IIIa. The monoadsorbed alkane can either desorb (rate constant k_3) or form diadsorbed alkane (k_{-2}). In Fig. 12 k_{-2}/k_3 is small on W but large on Ir.

With heptane, the most exchanged species is the perdeutero, but with 3,3-dimethylpentane and neopentane, it is

$$CH_3CH_2-\overset{\displaystyle CH_3}{\underset{\displaystyle CH_3}{\overset{|}{\underset{|}{C}}}}-CD_2CD_3 \quad \text{and} \quad CH_3-\overset{\displaystyle CH_3}{\underset{\displaystyle CH_3}{\overset{|}{\underset{|}{C}}}}-CH_2D$$

Thus, the reaction $*CHR-CH_2-CH_2R' + 2* = *CHR-CH_2-CHR'* + H*$ does not occur in the presence of hydrogen. However, if one starts with a reactant of higher free energy, 1,3-diadsorbed alkanes can become major intermediates even in the presence of hydrogen, for example in the hydrogenolysis of cyclopropane to propane. Presumably, 1,4-diadsorbed alkanes are intermediates in the more difficult hydrogenolysis of cyclobutanes.

Rates of isotopic exchange of alkanes are roughly of the form,

$$\text{rate} = KP_{alkane}{}^{a}P_{H_2}{}^{-b}$$

where a and $-b$ lie between about 0.5 and 1.0. Thus hydrogen is strongly adsorbed and covers most of the surface at lower temperatures. Similar kinetics and conclusions apply to the hydrogenolysis of cyclopropanes. In the hydrogenation of olefins, however, kinetics indicates that olefin is much the more strongly adsorbed. In confirmation, olefins inhibit the hydrogen-deuterium exchange reaction.

Wide isotopic distributions of alkane result from the addition of deuterium to olefin (Fig. 11). Thus the adsorbate-gas reaction $2H* + C_2H_4(g) = 2* + C_2H_6(g)$, Vb, is unimportant. However, the initially formed alkane is less highly exchanged than that in alkane exchange in the absence of olefin, for two reasons. The presence of olefin greatly reduces surface coverage by deuterium and it inhibits exchange between $*$-H and $D_2(g)$, a reaction which is very fast in alkane exchange. We have already seen that exchanged and isomerized olefin can desorb from the surface during olefin hydrogenation. These three factors all make isotopic tracer experiments harder to interpret in the reaction between deuterium and olefins than in alkane exchange. However, the olefin deuterogenation reveals additional differences among metallic catalysts — for example, that the rate of formation of isomeric olefins relative to that of hydrogenation is large on Pd and small on Pt.

A number of exchange experiments indicate that, if the diadsorbed species of Fig. 11 is 1,2-diadsorbed alkane, its conformation must be eclipsed and not staggered. For example, adamantane, shown in Fig. 13, exchanges only one deuterium atom in one period of residence on the catalyst. Any 1,2-

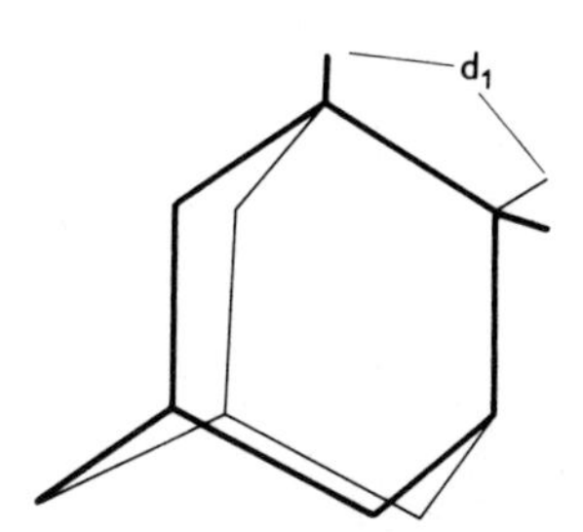

Fig. 13. Adamantane.

diadsorbed adamantane could only be staggered. Two eclipsed hydrogen atoms on one of the ethylene bridges of bicycloheptane (Fig. 14) exchange in one period, but no more since this would require either 1,3- or staggered 1,2-diadsorbed intermediates as may be seen in Fig. 14.

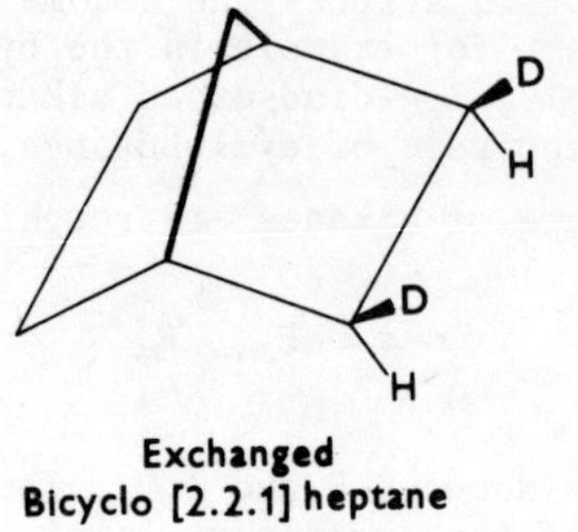

Exchanged
Bicyclo [2.2.1] heptane

Fig. 14.

The recent suggestion that the diadsorbed intermediate is a π-complexed olefin (**E** in Fig. 1) accords just as well with these data since the total difference in the geometry between **E** and **B** is not very large. However, results of isotopic exchange on Pd of other cyclic alkanes indicate that a planar π-complex cannot be involved. If a π-complex is an intermediate its geometry must approach that of **B** in Fig. 1. The exact details of the binding of the intermediate **B**-**E** are not resolved.

Isotopic exchange between cyclopentane and deuterium at lower temperatures (first studied by Kemball) has been of particular interest. An example for Pd is shown in Fig. 15. Alternation between monoadsorbed and eclipsed

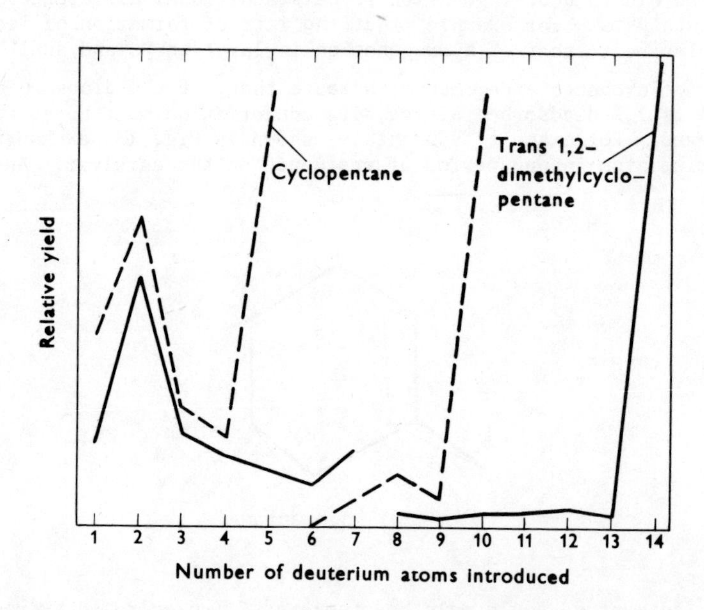

Fig. 15. Isotopic exchange between deuterium and cyclopentane at 40 °C and *trans*-dimethylcyclopentane at 60 °C, on Pd/Al_2O_3 catalyst.

diadsorbed would exchange but one side of a cyclopentane ring and give d_5 as the maximally exchanged species. Some additional process occurs to exchange both sides of the ring to give d_{10}. The exchange pattern for *trans*-1,2-dimethylcyclopentane is also shown in the Figure. Each methyl group exchanges with the set of hydrogen atoms on the other side of the ring via the diadsorbed intermediate;

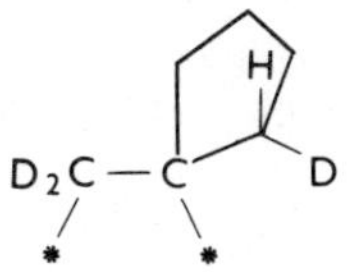

H

D_2C — C D

* *

In addition, epimerization occurs to form *cis*-1,2-dimethylcyclopentane, which is almost completely perdeutero. A number of polymethylcyclopentanes have been studied and all give patterns according with predictions. The fact that cyclooctane does not show a break between $C_8H_8D_8$ and $C_8H_7D_9$ also agrees with the proposed mechanism. *Trans*-eclipsed diadsorbed cyclooctane is readily possible and it serves to transfer adsorption from one side of the ring to the other.

What is the mechanism by which more than five hydrogen atoms exchange in cyclopentane and by which *cis*- and *trans*-dimethylcyclopentane are interconverted, and by which (+)3-methylhexane is racemized during exchange? The answer is not yet clear but probably it involves, at least in some cases, a process in which a diadsorbed species rolls over on the surface. Nor are we clear as to the exact nature of the transition states in the various elementary steps which have been suggested. It will be difficult to make really major advances in the last problem until we know more about the detailed nature of the surfaces and binding. A similar conclusion applies to our poor understanding of the nature of the differences among the metallic catalysts.

Reactions on Oxides

We know something about reactions catalysed by such oxides as Cr_2O_3, ZnO (see the following article by Kokes), TiO_2 (Kemball), and Al_2O_3 (Hall and Hightower). Reactions on Cr_2O_3 may be taken to exemplify some of the reactions catalysed by this group. Cr_2O_3 originally attracted attention because of its surprising ability to convert heptane to toluene. A Cr_2O_3 catalyst — for example a high-area gel — needs to be activated by treatment at high temperatures before it acquires catalytic activity for the hydrogenation of olefins or for the exchange of alkanes. It seems to be generally agreed that the activation involves elimination of water and the formation of surface Cr atoms which are coordinatively unsaturated. It may also involve reduction of Cr(III) to Cr(II) under some conditions but this point is unsettled.

In addition, certain forms of Cr_2O_3 have substantial capacity for catalysing double-bond migration in such olefins as 1-pentene. There is evidence that heterolytic dissociative adsorption to $(allyl)^-Cr^{3+} + OH^-$ is involved as an intermediate. Thus $CH_2{=}CH{-}CH_2{-}R \xrightarrow{-H^+} [CH_2{\cdots}CH{\cdots}CH{-}R]^- \xrightarrow{+H^+} CH_3{-}CH{=}CH{-}R$ results in double-bond migration.

Activated Cr_2O_3 adsorbs hydrogen and it rapidly catalyses the hydrogen-deuterium exchange reaction and the hydrogenation of olefins at room temperatures. In contrast to the results on metal catalysts, the product of the addition of deuterium to ethylene is exclusively CH_2DCH_2D. With *cis*-2-butene the product is *meso*-2,3 dideuterobutane.

The mechanism in Fig. 16 seems a reasonable one, and, in particular, close homogeneous analogies to it have developed. Ethylene is adsorbed by reactiv adsorption. Perhaps it first adsorbs as a π-complex and then reacts to mono adsorbed ethane. This latter formulation emphasizes the resemblance of the reaction to the insertion reaction of organometallic chemistry. In any case either no diadsorbed alkane exists on Cr_2O_3 as in our formulation, or if one does exist, its formation from monoadsorbed alkane, IIIa in Fig. 2, is very slow. The mechanism predicts that isotopic exchange between alkane and deuterium should exchange only one hydrogen atom as is indeed observed. The relative rates of exchange of alkanes suggest that the alkyl

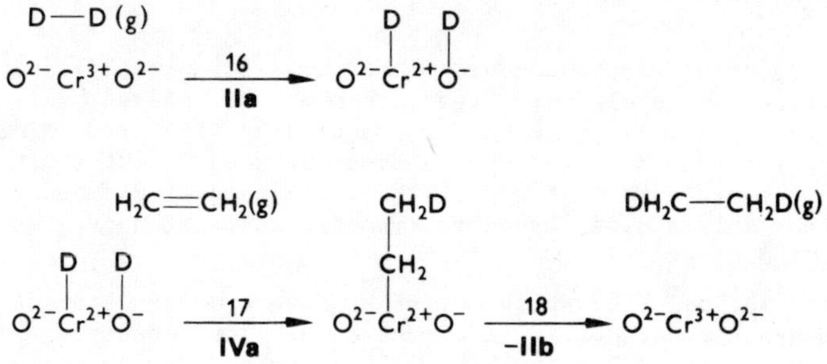

$$\text{D—D(g)} \quad O^{2-}Cr^{3+}O^{2-} \xrightarrow[\text{IIa}]{16} O^{2-}\overset{\text{D}}{Cr}{}^{2+}\overset{\text{D}}{O}{}^{-}$$

$$\text{H}_2\text{C=CH}_2\text{(g)} \quad O^{2-}\overset{\text{D}}{Cr}{}^{2+}\overset{\text{D}}{O}{}^{-} \xrightarrow[\text{IVa}]{17} O^{2-}\overset{\text{CH}_2\text{D–CH}_2}{Cr}{}^{2+}O^{-} \xrightarrow[\text{-IIb}]{18} O^{2-}Cr^{3+}O^{2-} + \text{DH}_2\text{C—CH}_2\text{D(g)}$$

Fig. 16. The mechanism for the hydrogenation of olefins on Cr_2O_3.

group has a small amount of carbanion character in the transition state of reaction 18 in Fig. 16. In contrast, on metallic catalysts, the transition state probably has a small amount of carbonium ion character since neopentane exchanges much more slowly than pentane.

Nonmetallic catalysts of this nature are of particular interest as a corrective to getting all of our opinions on the mechanisms of hydrogenation reactions from metallic catalysts and because of the interrelation of reactions on nonmetallic catalysts with homogeneous reactions catalysed by complexes of transition metals. The understanding of the feature of activation is a challenge, but the phenomenon of activation is of considerable assistance in study of the chemical mechanism.

Reactions on Silica-Alumina

One other type of catalyst, silica-alumina, has been widely studied with respect to hydrocarbon reactions. Amorphous silica-aluminas and certain crys talline silica-aluminas (of which synthetic zeolites are of current interest) exhibit a characteristic set of reactions which closely resemble those promoted by such strong acids as sulfuric acid and aluminium chloride. Further, as tested by reactions of indicators, surface sites on silica-alumina encompass a range of acidities including the very strong. It is clear that the active sites constitute a rather small portion of the surface of the amorphous silica-aluminas. C.L. Thomas was the first to conclude that the reactions on silica-alumina are of a carbonium-ion type. One can readily write carbonium-ion mechanisms for the following reactions which occur on these catalysts:

$$C_6H_5CH(CH_3)_2 \longrightarrow \text{benzene} + CH_3CH{=}CH_2 \quad 350°C$$

$$\text{isobutane} + D_2O \longrightarrow \text{multiply exchanged isobutanes} \quad 150°C$$

$$\text{dodecane} \longrightarrow \text{branched octanes} + \text{isobutane} \quad 450°C$$

(The temperatures indicated are those appropriate for amorphous silica-aluminas).

It is the last reaction, the cracking reaction, which originally attracted attention to these materials and which is run on the largest tonnage scale of almost any chemical reaction. Of course, the specific reaction listed is just one of myriads of similar ones.

There are three main unsolved problems in this area:

First, what is the mechanism whereby a carbonium ion is first formed from an alkane? Given the carbonium ion, standard carbonium-ion mecharisms can be applied.

Secondly, what is the exact chemical structure of the active site?

Thirdly, what is the nature of the binding of the carbonium ion to the sites? The energy of a free carbonium ion is so high that considerable interaction is needed to stabilize it, much as in solvation of a carbonium ion in a medium of high dielectric constant.

Dual-Functional Catalysts

We know something about the mechanism of reactions on catalysts consisting of Pt deposited on acidic alumina. These catalysts are widely used for *reforming* gasoline – for improving the quality of gasoline principally with respect to its octane number. Three reactions occur at about 450 °C or above in the presence of excess hydrogen:

(1) Isomerization of straight-chain to branched-chain alkanes, isomerization of methylcyclopentanes to cyclohexanes.

(2) Dehydrogenation of alkylcyclohexanes to alkylbenzenes.

(3) Dehydrocyclization of heptane to toluene.

The catalyst combines the functions of an acidic and a hydrogenation catalyst. The latter function produces dehydrogenation of alkylcyclohexanes. It also appears to be involved in formation of carbonium ions via dehydrogenating alkanes to olefins. The olefins migrate to acidic sites and react there to form carbonium ions, which lead to isomerization reactions. Even though equilibrium considerations restrict olefin concentrations to small values, olefin may migrate from metallic to acidic sites through the gas phase since Weisz has shown that intimate mixtures of acidic and hydrogenation catalysts behave like the dual-functional catalysts themselves. The mechanism of dehydrocyclization is obscure as are many details of the isomerization reactions.

ACKNOWLEDGMENT

We thank the American Chemical Society for permission to reproduce sections of an article by R.L. Burwell, Jr., in *Chemical Engineering News*, August 22, 1966.

REFERENCES

General

BOND, G.C. (1962). *Catalysis by Metals*, (Academic Press, London).
THOMSON, S.J. and WEBB, G. (1968). *Heterogeneous Catalysis*, (Wiley, New York).
ANDERSON, R.B. (1968). *Experimental Methods in Catalytic Research*, (Academic Press, New York).

Surface Chemical Physics

SOMORJAI, G.A. (1972). *Principles of Surface Chemistry*, (Prentic Hall, Englewood Cliffs, N.J.).

Chemisorption

HAYWARD, D.O. and TRAPNELL, B.M.W. (1964). *Chemisorption*, (Butterworths, London).

Physisorption and Texture

GREGG, S.J. and SING, K.S.W. (1967). *Adsorption, Surface Area and Porosity*, (Academic Press, London).

Isotopic Exchange in Heterogeneous Catalysis

BURWELL, R.L., Jr. (1972). *Catalysis Reviews*, 7, 25.
EMMETT, P.H. (1972). *Catalysis Reviews*, 7, 1.

Stereochemistry in Heterogeneous Catalysis

SIEGEL, S. (1966). *Adv. Catalysis*, 16, 123.

Reactions on Chromia

BURWELL, R.L., Jr.,, HALLER, G.L., TAYLOR, K.C. and READ, J.F. (1969). *Adv. Catalysis*, 20, 1.

Catalysis by Solid Acids and Bases

TANABE, K. (1970). *Solid Acids and Bases, Their Catalytic Properties*, (Academic Press, New York).

THE NATURE OF ACTIVE SITES

R. J. Kokes
Department of Chemistry, The Johns Hopkins University, Baltimore, Maryland 21218, USA

Catalysts function by formation of intermediates from reactants which subsequently decompose to yield products. Usually, catalysts have some degree of valence unsaturation so that they can form bonds to the intermediates, but these bonds must also be readily ruptured so that intermediates can readily form products. For most heterogeneous catalysts, reactions are limited to the atoms on the surface, and the valence unsaturation stems from the 'dangling bonds' of the surface atoms, which possess a lower coordination number than the sub-surface atoms. If, however, the solid catalyst consists of well-formed crystallites, not all surface atoms have the same coordination number, and, hence, the degree of unsaturation may be different for different atoms. For example, the coordination number of a Pt atom in a 111 surface plane is 9 compared to 12 for the bulk, but the coordination number of a Pt atom in the 110 surface plane is 7. If one includes atoms along edges of intersecting planes or at steps, even smaller coordination numbers are found. In general, the lower the coordination number, the higher the valence unsaturation attributed to the surface atoms. Thus, these different atoms could well be expected to exhibit different activity in the overall catalysis effected by a single crystal. Langmuir[1] reached the same conclusion fifty years ago. In his discussion of activation of Pt catalysts, he concluded: 'The surface thus becomes a composite, and there is then a relatively small fraction of the surface at which reaction occurs with extreme rapidity while over the larger part of the surface, it takes place at a very slow rate'. It was immediately recognized[2] that such a picture gave ready explanation of the fact that trace impurities (in amounts that could react with only a fraction of the surface) could reduce the catalytic activity by orders of magnitude. To explain this we need only assume that the poison reacts selectively with the most active sites. It was not until Taylor's classic paper in 1925[3], however, that the concept of active centers could be regarded as clearly developed. Today, this concept, properly used, must be regarded as indispensible in general discussions of heterogeneous catalysis[1].

There is little doubt that active centers exist; the question is, what is their nature. If we considered only high index planes of perfect crystals and the edges they form, the possibilities would be somewhat limited. Today, however, we know that we must also consider point defects, line defects, edge defects, electronic defects and combinations thereof. Add to these the possibility of surface reconstruction to form still other structures and the list of conceivable active centers approaches infinity. Thus, it seems pointless in this brief presentation to try to catalog the variety of possible active centers. Instead, what we shall do is select two examples of catalysts, Pt and ZnO, and discuss for these a few of the experimental observations that shed some light on the nature of active centers. In so doing, we shall try to add to the picture of the active centers only those features required by the experiments. Such an approach may not lead to a very graphic picture of the active sites, but it is a useful exercise insofar as it emphasizes how limited the experimental basis is for understanding the structure of the active sites.

PLATINUM CATALYSTS

It is the surface atoms that are effective in heterogeneous catalysts; hence, effective utilization of the available Pt requires that most of it be in the form of small crystallites so that there is a large surface to volume ratio. Pure Pt catalysts can be prepared with very small crystallite size. At elevated temperatures, however, such catalysts readily undergo sintering and the effective crystallite size increases. Accordingly, most practical catalysts are made by dispersing Pt on a high area 'inert' support such as silica gel, alumina, or carbon. Such catalysts, perhaps because of the lack of crystallite-crystallite contact, are, in general, much more resistant to sintering.

Consider a series of supported platinum catalysts spanning a large range of crystallite sizes. For well-formed crystallites the fraction of surface atoms with the low coordination number characteristic of intersecting planes should increase as the crystallite size decreases. If those surface atoms with low coordination number are especially effective in catalysis, one would expect the reaction rate *per surface metal atom* to increase for the better dispersed samples. Such studies have been carried out. For a number of reactions it is found that the rate per surface atom is independent of the crystallite size[4]. The simplest interpretation of this result is that catalysis of this class of reactions is effected by all surface atoms and that the effectiveness of a given site is not very sensitive to its structure. Accordingly, Boudart[4] has suggested that such reactions be called *facile* or *structure-insensitive* reactions.

Table I lists a series of known facile reactions for Pt . The list is quite diverse in terms of the chemistry that must be involved. If we follow

TABLE I

Facile Reactions over Pt†

$$SO_2 + \tfrac{1}{2}O_2 \rightarrow SO_3$$

$$H_2 + \tfrac{1}{2}O_2 \rightarrow H_2O$$

$$H_2 + D_2 \rightarrow 2HD$$

$$\text{cyclo } C_6H_{10} + H_2 \rightarrow C_6H_{12}$$

$$\text{cyclo } C_6H_{12} + H_2 \rightarrow C_6H_{14}$$

$$1\text{-}C_6H_{12} + H_2 \rightarrow C_6H_{14}$$

$$CH_3\text{-}\underset{\displaystyle OH}{\underset{|}{CH}}\text{-}CH_3 \rightarrow CH_3\text{-}\underset{\displaystyle O}{\underset{\|}{C}}\text{-}CH_3 + H_2$$

$$\text{cyclo } C_5H_{10} + H_2 \rightarrow C_5H_{12}$$

$$C_6H_6 + 3H_2 \rightarrow \text{cyclo } C_6H_{12}$$

$$\text{cyclo } C_3H_6 + H_2 \rightarrow C_3H_8$$

† Based on Table I in ref. 5.

the line of reasoning in the preceding paragraph, we would assume that all surface atoms are functioning (although perhaps not equally well) as active centers. Since surface atoms differ quite significantly in coordination number and the fraction of sites with differing coordinations should depend on crystallite size, the fact that these reactions are facile suggests that the immediate environment, i.e. the structure of these active sites, has little to do with their reactivity.

(The above conclusion is hardly unequivocal. One can assume, for example, that the fraction of the surface atoms that are very active sites does not depend critically on crystallite size. Alternatively, one can assume that a softening of the effect of a changing distribution of sites is a consequence of statistical averaging imposed on a rectilinear free energy relation[6]. In such a view the overall catalysis from a distribution of sites differing in activity is effected by a broad band of sites in the distribution. Both the above views are attractive because they offer a straightforward explanation for the fact that poisoning effects may depend on particle size[4]. In detail, however, these views are unlikely. It is difficult to believe that the distribution of surface structures is *not* a function of crystallite size and statistical averaging cannot wipe out the effect of a changing distributin of sites unless the activities of the sites are relatively independent of their structures).

Not all surface processes are facile; some may depend on the detailed structure of the surface atom. Molecular nitrogen adsorption on metals furnishes one of the most striking examples of such effects. Van Hardeveld and Van Montfoort measured heats of adsorption and observed the IR spectrum for this species on a number of supported metals (including Pt) as a function of crystallite size[7]. They concluded that the molecular adsorption became more extensive for smaller crystallites and argued that this adsorption occurred only on a special kind of site designated as a B_5 site. (A B_5 site is one in which the coordination number of an adsorbed molecule with catalyst atoms is 5. In their model the ratio of B_5 sites to the total number of atoms increases with decreasing crystallite size and reaches a maximum at a crystal diameter of about 25 Å). Some reactions can also be structure-sensitive. Boudart *et al.*[5] studied the isomerization and hydrogenolysis of neopentane over a graded series of Pt catalysts. The following reactions are observed:

$$C(CH_3)_4 + H_2 \longrightarrow \begin{array}{ccccccccc} & & H & & H & & H & & H \\ & & | & & | & & | & & | \\ H & - & C & - & C & - & C & - & C - H + H_2 \\ & & | & & | & & | & & | \\ & & H & & H & H - & C & - H & H \\ & & & & & & | & & \\ & & & & & & H & & \end{array}$$

$$\longrightarrow \begin{array}{ccccccc} & & H & & H & & H \\ & & | & & | & & | \\ H & - & C & - & C & - & C - H + CH_4 \\ & & | & & | & & | \\ & & H & H - & C & - H & H \\ & & & & | & & \\ & & & & H & & \end{array}$$

The relative reaction rate (selectivity) ranges from a ratio of 0.55 to 27 depending on sample preparation. They believe that the first reaction is enhanced when the molecule can adsorb on a triplet of Pt atoms. Reactions such as these, which are structure-sensitive, have been termed *demanding*.

Studies of single crystals by *low-energy electron diffraction* (LEED) and *Auger electron spectroscopy* (AES) have provided rather dramatic and detailed evidence for the existence of surface atoms with very different reactivities[8]. These studies suggest that flat 111 surfaces are quite unreactive. If a cut is made a few degrees off the 111 surface, however, a stable surface is created with 111 terraces and regularly spaced 100 steps. This surface is much more reactive. By way of illustration, adsorption of hydrogen and oxygen on the unstepped 111 surface occurs only for 10^{-2} and 10^{-6} of the collisions with the surface, respectively. Both gases adsorb much more readily on the stepped surface. There is some evidence, also, that adsorption starts at the step and spreads across the terraces. Observations of the conversion of heptane to toluene on the various faces of single crystals have been made on the LEED apparatus. The rate is faster initially and self poisoning is less evident on the stepped surface than on flat surfaces.

The results of the LEED experiments certainly provide strong evidence that under the condition of these experiments, the structure of surface sites plays a major role in their activity. In fact, if one judged by the LEED experiments alone, one would conclude that the surface is so reactive that hydrocarbons are initially carbonized and a working catalyst is carbonized Pt. It should be emphasized, however, that the conditions under which LEED observations are made (e.g. pressures below 10^{-4} torr) are far different from those used in the usual catalytic processes. Often, for practical catalysts hydrogen is required to prevent activity loss. It is possible that, in these rather high pressures of hydrogen, the surface chemistry characteristic of LEED experiments may be dramatically altered. In other words, extrapolation of the LEED results to those for the usual catalyst under its usual working conditions may be unjustified.

ZINC OXIDE[9]

If one exposes an activated sample of zinc oxide to hydrogen at room temperature, there is a rapid adsorption of hydrogen followed by a slower process. To a first approximation this hydrogen adsorption can be separated into two types. The first, type I, is rapid and reversible; the second, type II, is irreversible and, although some of this occurs rapidly, it is the major component of the slow hydrogen chemisorption. Tracer experiments, at or near room temperature, have shown that type II hydrogen is inert in ethylene hydrogenation and hydrogen-deuterium exchange, but that type I hydrogen participates in both of these reactions[9,10]. Selective adsorption experiments show that the sites for type I adsorption constitute only a fraction of the surface (5 to 10%) and that when these are blocked out by the selective adsorption of water, reaction is poisoned. Thus it appears that the sites for type I hydrogen adsorption are those responsible for hydrogen activation and, hence, these are the active sites for these reactions.

Fig. 1 shows the IR spectrum of hydrogen adsorbed on zinc oxide. This spectrum, characterized by strong bands at 3489 cm^{-1} and 1709 cm^{-1}, was first reported by Eischens *et al.*[11] who assigned these bands to an OH and ZnH surface species. Coupled IR and adsorption experiments show that these bands stem from the type I chemisorption. Thus, it is reasonable to suppose that type I hydrogen adsorption can be represented as follows:

$$H_2 + Zn—O \rightleftharpoons H—Zn—O—H$$

where Zn-O represents an active site. Only 5 to 10% of the surface contains such active sites.

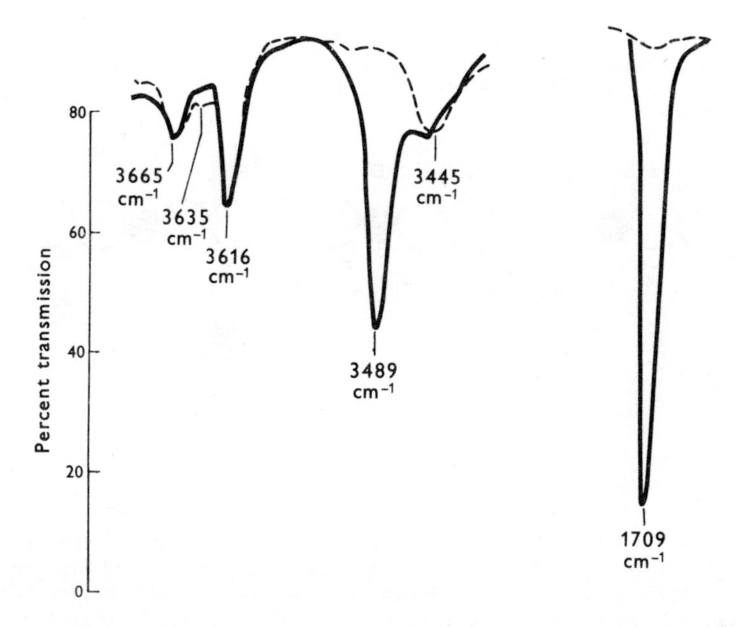

Fig. 1. Spectrum of hydrogen on zinc oxide. The broken line is the spectrum for the degassed catalyst.

The IR spectrum of the dissociatively adsorbed hydrogen shows that both cation and anion are involved in the active site but does not specify the relative positions of these two sites. Recent studies of isotope effects shed further light on the nature of these sites. At room temperature the adsorption-desorption equilibrium for type I hydrogen is rapid; at lower temperatures (below -40 °C) adsorption is irreversible. Consider the adsorption of hydrogen deuteride on Zn-O pair sites. We can imagine two processes:

$$\mathrm{H{-}D + ZnO \longrightarrow \overset{H}{\overset{|}{Zn}}{-}\overset{D}{\overset{|}{O}}} \quad \mathrm{(HD)}$$

$$\longrightarrow \mathrm{\overset{D}{\overset{|}{Zn}}{-}\overset{H}{\overset{|}{O}}} \quad \mathrm{(DH)}$$

These two species can be distinguished by examination of the IR spectrum[12]. Observations of adsorption at -195 °C reveal that the kinetic isotope effect is pronounced; the rate of formation of the DH is faster by nearly two orders of magnitude. The thermodynamic isotope effect (at room temperature) favors HD by a factor of three. If we adsorb hydrogen deuteride at -195 °C and warm the catalyst to room temperature, the preferred form changes from DH to HD; if we recool the catalyst to -195 °C, the equilibrium preference for HD becomes frozen in. These results suggest that not only are the sites few in number, covering 5 to 10% of the surface, but that they are isolated metal-oxide pairs. Adsorption occurs directly on these sites at low temperatures and equilibrium is achieved by site-to-site migration or interchange at higher temperatures.

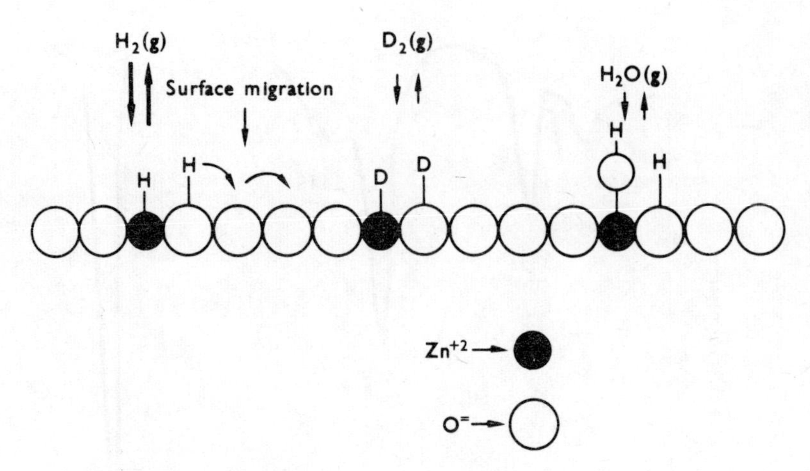

Fig. 2. Schematic picture of the active sites.

Fig. 2 represents a schematic picture of such isolated pair sites on zinc oxide. If type I hydrogen is the only activated form of hydrogen, site-to-site migration is required to accomodate hydrogen-deuterium exchange. Adso tion studies and comparative rate studies of ethylene hydrogenation with is topic mixtures suggest adsorption-desorption is faster than site-to-site migration. Since para hydrogen conversion probably requires only adsorptio desorption, one would expect it to be faster than hydrogen-deuterium exchan experiments in this laboratory conform to this expectation. The picture in Fig. 2 includes all of the features necessary to explain the experimental data but says little about the origin of these sites. At present, all we wish to specify is that these sites do not stem from oxygen deficiency. A catalyst sample pretreated at high temperature with oxygen and cooled in oxygen to room temperature yields essentially the same results as pretreatment *in vacuo* at high temperatures.

Let us re-examine the picture in Fig. 2 in more detail. Water poisoning of these sites can be pictured as follows:

$$H_2O + Zn^{2+}—O^{2-} \longrightarrow \overset{OH^-}{\underset{|}{Zn^{2+}}}—\overset{H^+}{\underset{|}{O^{2-}}}$$

There is, in fact, some sketchy evidence that new hydroxyl bands appear on water poisoning. In this view, water poisoning involves heterolytic cleavage A parallel view of hydrogen adsorption would suggest that this occurs as follows:

$$H—H + Zn^{2+}—O^{2-} \longrightarrow \overset{H^-}{\underset{|}{Zn^{2+}}}—\overset{H^+}{\underset{|}{O^{2-}}}$$

Some justification for this formalism can be found in the character of the spectrum in Fig. 1. The Zn-H band is roughly as intense as the OH band. All other things being equal (which they seldom are) the band intensities for such stretching frequencies should reflect the polarity of the bond.

The electronegativity difference of zinc and hydrogen is far less than that for oxygen and hydrogen; hence, the relatively high intensity of the Zn-H band suggests that the hydrogen bound to the zinc has an unexpectedly high degree of hydridic character. Thus, one of the characteristics of the Zn-O pair site would appear to be its ability to promote heterolytic cleavage.

One of the proposed mechanisms of olefin isomerization over zinc oxide (and related metal-oxide hydrogenation catalysts) involves the formation of allylic species. Studies of the IR adsorption of propylene and other olefins indicate that adsorption on the catalyst occurs by rupture of the allylic C-H bond as follows:

$$CH_3—CH{=}CH_2 + Zn—O \longrightarrow \begin{array}{cc} CH_2{=\!=}CH{=\!=}CH_2 & H \\ & | \\ Zn & —O \end{array}$$

The π-allyl species thus formed is the intermediate in olefin isomerization of labelled propylenes. The pair sites involved in this dissociative adsorption include those needed for type I adsorption; preadsorption of propylene prevents the formation of type I hydrogen bands. The number of such sites available for π-allyl formation is, however, greater than the number utilized for type I hydrogen chemisorption. This could be a reflection of the energetic heterogeneity of these pair sites. Given a distribution of site energies, the more strongly bound π-allyl may be able to utilize more low energy sites than the relatively loosely held hydrogen.

If we accept the supposition that these pair sites function by heterolytic cleavage, there are some mechanistic consequences. On this basis the π-allyl should have considerable anionic character. Accordingly, one might expect the isomerization reaction to resemble those in homogeneous base catalysts wherein allyl carbanions are the presumed intermediates. Isomerization of butene-1 by homogeneous base catalysts yields very high initial *cis* to *trans* ratios (i.e. 10 to 40) in the butene-2 products whereas isomerization by other pathways generally yields *cis/trans* ratios of the order of unity. Over zinc oxide the initial *cis/trans* ratio is about ten, a value closer to that observed for homogeneous base-catalysed reactions than to that observed for isomerization over metal hydrogenation or acidic-oxide catalysts.

Coupled IR and adsorption experiments with a variety of hydrocarbons suggest that the pair sites on zinc oxide possess a rather high electric field that can heterolytically cleave many acidic carbon-hydrogen bonds. As such, these sites may serve as active centers for preferential molecular adsorption (by polarization) as well as sites for heterolytic dissociation. Molecular hydrogen adsorption furnishes an example of such adsorption. Fig. 3 shows the IR bands observed when zinc oxide is exposed to hydrogen, hydrogen deuteride, and deuterium at low temperatures. Table II lists the positions of these bands along with those reported for the fundamental vibrations for the gaseous molecular species[13]. There can be no doubt that the observed bands are due to molecular species with the frequencies shifted down due to interactions with the surface. In the gas phase, IR transitions due to vibrations of homonuclear diatomic molecules are forbidden, but it has been recognized for some time that such transitions can be induced by strong electric fields[14]. Sheppard and Yates[15] have observed such transitions for physically adsorbed hydrogen, but the intensity of the bands we observe[16] are more intense by a factor of 65. If we utilize their procedure (admittedly crude) for estimating the field required to induce our observed bands for hydrogen, we come to the conclusion that the effective field should be of the order of 1 V/Å.

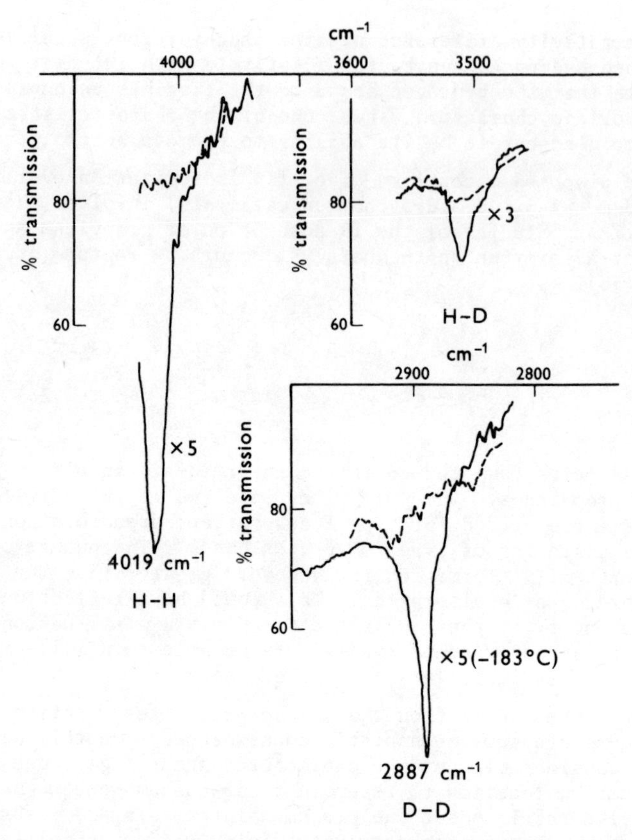

Fig. 3. Spectrum of molecular H_2, HD, and D_2 adsorbed on zinc oxide. The broken line is the spectrum of the degassed catalyst. For H_2 and D_2 the temperature was -195 °C.

TABLE II

Vibration Frequencies for Molecular Hydrogen

Species	ν (gas)(cm^{-1})[13]	ν (ads)(cm^{-1})	$\Delta\nu$ (cm^{-1})	$\frac{\Delta\nu}{\nu} \times 100$
H_2	4161	4019	142	3.41
HD	3627	3507	120	3.31
D_2	2990	2887	103	3.44
			Av	3.39 ± 0.5

Characteristics of the spectrum suggest that the interaction energy is much greater for the species on zinc oxide than it is on porous glass. Nevertheless, adsorption characteristics suggest that the binding is still rather weak; for example, the molecular bands disappear after brief evacuation at -195 °C. Despite the low heat of binding this molecular adsorption appears to occur on the same sites that are responsible for type I adsorption. If the catalyst is poisoned with water, the bands for both type I and molecular hydrogen no longer appear. As the catalyst is regenerated by removal of adsorbed water in stages, both bands reappear and grow in concert. These and other experiments suggest that the same active sites are utilized simultaneously for the dissociative type I and molecular adsorption. Thus, despite the fact that molecular hydrogen adsorption is weak, it appears to occur only on unpoisoned type I sites.

(It seems clear that this molecular hydrogen adsorption may play a role in reactions involving hydrogen activation. In fact, the mechanism proposed by Tamaru *et al.*[10] for hydrogen-deuterium postulates such a molecular species. Investigations aimed at assessing its role (if any) in such reactions are now underway, but, at the present time, we have no evidence that it plays a role in chemical reactions).

In conclusion, these experimental results suggest that activated zinc oxide contains a limited number of non-interacting, metal-oxide, pair sites. These sites have a rather high effective field associated with them and are capable of heterolytic dissociation of chemical bonds. They may also serve as specific sites for weak molecular adsorption (by polarization). Beyond this point, the structure of these sites is a matter of speculation, but they do not appear to have their origins in non-stoichiometry generated by oxygen deficiency.

ACKNOWLEDGEMENT

Acknowledgment is made to the donors of the Petroleum Research Fund, administered by the American Chemical Society, for support of part of this research. This research was also aided by funds from the National Science Foundation under grant GP 34034X.

REFERENCES

1. LANGMUIR, I. (1922). *Trans. Faraday Soc.*, 17, 607.
2. ARMSTRONG, E.F. and HILDITCH, T.P. (1922). *Trans. Faraday Soc.*, 17, 669.
3. TAYLOR, H.S. (1925). *Proc. Roy. Soc. (London)*, A108, 105.
4. BOUDART, M., ALDAG, A., BENSON, J.E., DOUGHARTY, N.A. and HARKINS, G.C. (1966). *J. Catalysis*, 6, 92.
5. BOUDART, M., ALDAG, A.W., PTAK, L.D. and BENSON, J.E. (1968). *J. Catalysis*, 11, 35; for a more complete discussion of supported metals, see: BOUDART, M. (1969). *Adv. Catalysis*, 20, 153.
6. BOUDART, M. (1968). *Kinetics of Chemical Processes*, (Prentice Hall, Englewood Cliffs, N.J.), Chapter 9.
7. VAN HARDEVELD, R. and VAN MONTFOORT, A. (1966). *Surface Sci.*, 4, 396.
8. SAMORJAI, G.A., JOYNER, R.W. and LANG, B. (1972). Presented at the *Royal Society Discussion on The Physics and Chemistry of Surfaces*, (London, England), May 23-25.
9. KOKES, R.J. and DENT, A.L. (1972). *Adv. Catalysis*, 22, 1.
10. NAITO, S. SHIMIZU, H., HAGIWARA, E., ONISHI, T. and TAMARU, K. (1971). *Trans. Faraday Soc.*, 67, 1519.
11. EISCHENS, R.P., PLISKIN, W.A. and LOW, M.J.D. (1962). *J. Catalysis*, 1, 180.
12. KOKES, R.J., DENT, A.L. CHANG, C.C. and DIXON, L.T. (1972). *J. Amer. Chem. Soc.*, 94, 4429.
13. HERZBERG, G. (1950). *Molecular Spectra and Molecular Structure. I. Spectra of Diatomic Molecules*, (Van Nostrand, New York), p. 533.

14. CONDON, E.U. (1932). *Phys. Rev.*, 41, 759; CRAWFORD, M.F. and DAGG, I.R. (1953). *Phys. Rev.*, 91, 1569.
15. SHEPPARD, N. and YATES, D.J.C. (1957). *Proc. Roy. Soc. (London)*, A238, 69.
16. CHANG, C.C. and KOKES, R.J. (1971). *J. Amer. Chem. Soc.*, 93, 7107.

SELECTIVITY AND POISONING

Charles Kemball
Department of Chemistry, University of Edinburgh,
King's Buildings, West Mains Road,
Edinburgh EH9 3JJ, Scotland

A number of terms will be given to describe some of the concepts of selectivity and poisoning in heterogeneous catalysis and then a few examples will be presented.

SELECTIVITY

Most heterogeneous catalysts are capable of accelerating a number of chemical reactions, e.g. a typical hydrogenation catalyst may bring about the hydrogenation of a variety of olefins or may convert an acetylene into either an olefin or an alkane. The term selectivity is used to describe the actual behavior of the catalyst when alternative reactions are possible.

Three main situations involving selectivity were defined by Wheeler[1] and are also reported in *Catalysis by Metals* by Bond[2].

Type I refers to the case with two different reactants, each competing for the surface and each forming products:

$$\begin{aligned} A &\rightarrow B, \\ X &\rightarrow Y. \end{aligned} \tag{1}$$

A and X might be different olefins and B and Y the corresponding saturated hydrocarbons.

Type II involves a single main reactant but with two possible reaction paths and two sets of products, e.g. propan-2-ol undergoing dehydrogenation to form acetone and hydrogen or dehydration to form propylene and water:

$$A \begin{cases} \nearrow B \\ \searrow C. \end{cases} \tag{2}$$

Type III again involves one main reactant but consecutive products:

$$A \rightarrow B \rightarrow C, \tag{3}$$

which would be exemplified by acetylene forming ethylene and then ethane.

Mathematical expressions for the changes of concentration of the various substances with time can be derived for each of these main types of selectivity[1,2].

Definitions of selectivity are needed to describe observations with a given system involving selective catalysis. One possibility is fractional selectivity S_F given by:

$$S_F = \frac{\text{rate of formation of chosen product}}{\text{rate of formation of all products}} , \qquad (4)$$

and it follows that $0 \leqslant S_F \leqslant 1$, with values approaching unity when the chos product is produced selectively by the catalyst. Alternatively, the selectivity may be defined as a ratio:

$$S_R = \frac{\text{rate of formation of the chosen product}}{\text{rate of formation of all } \textit{other} \text{ products}} , \qquad (5)$$

and for this definition, it follows that $0 \leqslant S_R \leqslant \infty$.

It is sometimes helpful to distinguish two factors which play a part in controlling selectivity in heterogeneous catalysis — the thermodynamic factc and the mechanistic factor[3]. In Type I selectivity, competition between the two reactants A and X for the surface sites is important and if the rates of adsorption and desorption of these substances are rapid compared with the catalytic steps, the relative amounts of the adsorbed species will be determined by equilibrium considerations which constitute the thermodynamic factc Of course, if the rate of adsorption of one reactant X is slow compared with the rates of adsorption and desorption of A and with the rates of the catalytic steps, the concept of the thermodynamic factor is no longer relevant or useful. The mechanistic factor is important in Type II selectvity where there are alternative kinetic reaction paths, each with its rate of reaction associated with the mechanism involved. Both factors may play a role in Type III selectivity as may be seen by considering the extended reaction scheme:

$$\begin{array}{ccccc} A(g) & & B(g) & & C(g) \\ \downarrow\uparrow & & \downarrow\uparrow & & \uparrow \\ A(ads) & \rightarrow & B(ads) & \rightarrow & C(ads). \end{array} \qquad (6)$$

The fate of B(ads), i.e. the desorption to form B in the gas phase or conversion to C(ads), is controlled by a mechanistic factor. Once B is formed in the gas phase competition between A and B for sites on the catalyst can develop and a thermodynamic factor will also contribute to the selectivity.

The term shape-selective catalysis has become important recently in con nection with the use of zeolites as catalysts. Molecules of suitable dimensions can enter or leave the pores of molecular sieves. Larger reactants cannot enter and larger product molecules, even if they can be formed, will not readily escape from the catalyst.

INHIBITION OR POISONING

The terms inhibition and poisoning have similar meanings except perhaps that poisoning may imply a more drastic effect on the rate of reaction than inhibition. Poisoning of a reaction may be observed even if no substances other than the reactants and products are present in the catalyst system. The products may compete strongly with the reactants for sites on the surface and thereby reduce the rate of reaction. In other cases, the reactant itself may form strongly absorbed species which inhibit the formation of the less strong absorbed species required for the reaction under study — this is known as self-poisoning — and is exemplified by coking up of a cracking catalyst.

When the reduction in the rate of a catalytic reaction occurs through the addition of a new substance (the poison) the action may be reversible, partly reversible, or irreversible. Which kind of effect is obtained depends on the strength of adsorption of the poison on the catalyst, and on the conditions (temperature, time, etc.) used to try and recover the activity of the catalyst after the supply of the poison has ceased.

SELECTIVE POISONING

The term is almost self-explanatory and refers to a situation where the presence of a poison causes a catalyst to operate selectively, or more selectively, than it would have without the poison. There are two main systems to be considered — catalysts with uniform sites or catalysts with a variety of active sites. When the catalyst has uniform sites the possibilities of selective poisoning are probably somewhat limited. The principal way in which the poison might increase selectivity would be by competing for the sites and contributing to the thermodynamic factor involved in the selectivity. For instance, it might be possible to increase the selectivity of a catalyst for the hydrogenation of acetylene to ethylene by adding a substance which was more strongly adsorbed than ethylene, did not undergo catalysis itself, and did not interfere appreciably with the adsorption of acetylene. It is not so easy to visualize selective poisoning operating through mechanistic factors unless the mechanism of the desired reaction requires a smaller number of surface sites than the reaction to be selectively poisoned.

With catalysts possessing a variety of catalytic sites (these are probably very much more common than catalysts with uniform sites), the possibilities of achieving selective poisoning are greater. In favorable cases, high selectivity for a particular reaction might be obtainable by finding poisons which were adsorbed selectively on the sites responsible for undesired side reactions.

EXAMPLES

(1) Hamilton and Burwell[4] showed that the hydrogenation of but-2-yne on Pd-on-alumina is not only highly selective but also highly stereoselective because *cis*-but-2-ene is the only detectable product until the acetylene is used up. Subsequently, the *cis*-but-2-ene reacts in two ways — hydrogenation to butane or isomerization to *trans*-but-2-ene followed by conversion to butane. The ratio of isomerization : hydrogenation is 2.6:1 giving another example of selectivity.

Dimethylsulfide was a selective poison; it reduced the rate of hydrogenation of the acetylene but it poisoned completely the further reactions of *cis*-but-2-ene.

Many other examples of selectivity involving reactions of alkynes and dienes on metal catalysts will be found in an article by Bond and Wells[5].

(2) The hydrogenation of ethylene is generally regarded as a simple non-selective type of catalytic reaction but this is not strictly correct[6]. The equilibrium constants for the following reactions at 298 K are:

$$C_2H_4 + H_2 \rightleftharpoons C_2H_6 \quad K_7 = 10^{17.7}\,atm^{-1} \tag{7}$$

$$C_2H_6 + H_2 \rightleftharpoons 2CH_4 \quad K_8 = 10^{12} \tag{8}$$

$$C_2H_4 + 2H_2 \rightleftharpoons 2CH_4 \quad K_9 = 10^{29.7} atm^{-1} \tag{9}$$

and, in the presence of excess hydrogen, the formation of methane and not ethane would be expected. Catalysts like Ni which are used for hydrogenation of olefins will carry out the hydrogenolysis reactions (8) and (9) but fortunately only at higher temperatures.

(3) The ability of W, Pd and other metals to act as catalysts for the exchange of saturated hydrocarbons with deuterium to form isotopically labelled molecules without causing breakdown to smaller molecules by reactions like (8) is another example of selectivity[7]. The detailed results of these exchange reactions provide a number of examples of selectivity involving kinetic factors. This will be illustrated for the exchange of ethane for which the reaction scheme is:

$$\begin{array}{c} C_2X_6(g) \\ \Big\Downarrow\Big\Uparrow k_2 \\ C_2X_5(ads) \underset{}{\overset{k_1}{\rightleftharpoons}} C_2X_4(ads) \end{array} \tag{10}$$

The initial pattern of products is determined by the relative sizes of the rate constants k_1 and k_2. On metals like W and Mo, $k_1 < k_2$ and the main initial product is C_2H_5D with decreasing amounts of the more highly exchanged species. On metals like Pd and Rh, $k_2 < k_1$ and several interconversions between adsorbed ethyl radicals and adsorbed ethylene molecules occur before the molecules return to the gas phase; the main initial product is C_2D_6 with decreasing amounts of the less highly exchanged speeies.

(4) Self-poisoning is sometimes found during the catalytic exchange of saturated hydrocarbons with deuterium on metals, particularly at higher temperatures; e.g. in results for the exchange of n-hexane with deuterium above 273 K on Rh or above 373 K on Pd films[8]. This effect is due to the formation of strongly adsorbed hydrocarbon species by further dissociation of the reversibly formed intermediates responsible for the exchange reactions. In terms of the scheme (equation (10)) it would correspond to further dissociation of the adsorbed ethyl and adsorbed ethylene species. A related inhibition or poisoning of exchange reactions of hydrocarbons on metal films may result from pre-adsorption of the reactant hydrocarbon on the catalyst; the adsorption of n-butane on W[9] at 423 K poisons the rate of exchange of butane and deuterium at 273 K by a factor of about 2000. The strongly adsorbed species formed in such experiments are probably intermediates for the hydrogenolysis of the hydrocarbon at higher temperatures[10].

(5) An interesting example of selective poisoning of γ-alumina catalysts has been reported recently[11]. The reactions studied were the exchange of the olefinic hydrogen atoms in but-1-ene with deuterium and the isomerization of the but-1-ene to form *cis*- and *trans*-but-2-ene. Small amounts of adsorbed carbon dioxide (less than 1.6×10^{13} molecules cm^{-2}) completely eliminated the exchange reaction but had little effect on the rate of the isomerization. The results support the idea that the two kinds of reaction of but-1-ene occur on different types of sites on the catalyst.

(6) The final example of selective catalysis concerns reactions of cyclohexylamine and hydrogen on an evaporated Pt film[12]. The expected

reaction was hydrogenolysis according to the equation:

$$C_6H_{11}NH_2 + H_2 \longrightarrow C_6H_{12} + NH_3 \qquad (11)$$

But the initial reaction at 407 K was:

$$C_6H_{11}NH_2 \longrightarrow C_6H_6 + NH_3 + 2H_2 \qquad (12)$$

i.e. an unexpected dehydrogenation to benzene despite the presence of excess hydrogen. After most of the amine had decomposed, the benzene was slowly converted to cyclohexane to give final products in accord with equation (11). The amine was a stronger poison than ammonia for the hydrogenation of benzene. The results suggested that considerable dissociation of C-H bonds in the amine had occurred on the metal surface before the C-N bond was ruptured.

REFERENCES

1. WHEELER, A. (1951). *Adv. Catalysis*, 3, 249.
2. BOND, G.C. (1962). *Catalysis by Metals*, (Academic Press, London and New York), pp. 131-134.
3. BOND, G.C., DOWDEN, D.A. and MacKENZIE, N. (1958). *Trans. Faraday Soc.*, 54, 1537.
4. HAMILTON, W.M. and BURWELL, R.L., Jr. (1961). *Actes du Deuxième Congrès International de Catalyse*, (Editions Technip, Paris), p. 987.
5. BOND, G.C. and WELLS, P.B. (1964). *Adv. Catalysis*, 15, 91.
6. KEMBALL, C. (1966). *Disc. Faraday Soc.*, 41, 190.
7. KEMBALL, C. (1959). *Adv. Catlysis*, 11, 223.
8. GAULT, F.G. and KEMBALL, C. (1961). *Trans. Faraday Soc.*, 57, 1781.
9. CORTÉS ARROYO, A. and KEMBALL, C. (1972). *J. Chem. Soc. Faraday Trans. I*, 68, 1029
10. KEMBALL, C. (1971). *Catalysis Rev.*, 5, 33.
11. ROSYNEK, M.P., SMITH, W.D. and HIGHTOWER, J.W. (1971). *J. Catalysis*, 23, 204.
12. MOSS, R.L. and KEMBALL, C. (1956). *Nature*, 178, 1069.

OXIDATION AND DEOXIDATION

S. J. Teichner
Université Claude Bernard, Lyon I,
and *Institut de Recherches*
Sur la Catalyse, Villeurbanne, France

CORRELATIONS IN OXIDATION AND IN DEOXIDATION REACTIONS ON METALLIC OXIDES

Surfaces of oxides used as catalysts at low temperatures (below about ¼ of the melting point in K) contain various defects (steps, vacancies, interstitials, etc.) which may be privileged sites for the catalytic reaction (Gravelle and Teichner, 1969). Satisfactory correlations may be then obtained between a defect surface structure of the catalyst and the catalytic activity. Indeed, the surface ionic mobility being restricted at such temperatures the defects are often frozen in. But when the catalytic reaction is studied at temperatures at which surface or bulk ionic mobility exists, the previous localized properties are smeared out by collective properties of the surface. The whole of the oxide surface takes part in the reaction at these temperatures, there being no localized areas (immobile areas) of high activity. It is for such conditions that the present Chapter attempts a search for correlations between the catalytic activity and some properties of oxide catalysts.

Complete Oxidation Reactions

Boreskov and coworkers (1968) pointed out that the catalytic activities of various metal oxides for complete oxidation reactions and the homomolecular oxygen isotopic exchange reaction are related to the bond energy of the oxides. According to their investigations the oxygen on the surface is bound with various strengths and the catalytic activity can be attributed to the weakly bound oxygen.

The problem then arises of determining the energy of the oxygen bond in oxides when their surfaces are largely saturated with oxygen. Boreskov and coworkers solved this problem by plotting (Fig. 1) the equilibrium oxygen pressure over oxides against $1/T$. The temperature range of this plot is 50 - 500 °C and the oxygen pressure range is from 10^{-4} to 10^{-1} torr. The oxides studied were: TiO_2, V_2O_5, Cr_2O_3, MnO_2, Fe_2O_3, NiO, CuO, ZnO. The slope of the first straight line for each oxide (NiO in Fig. 1) gives the bond energy of oxygen remaining at the surface after evacuation of the sample at 50 °C for 1 h. The slopes of the remaining lines denote the bond energies of the bound oxygen after removal of some oxygen of the surface layer. The energy q as a function of the amount S of the oxygen removed from the surface is shown in Fig. 2 for a few oxides. Extrapolation to the origin gives the energy q_0 for $S = 0$. No correlation was found between these values of q_0 and the thermodynamic heats of formation of the oxides from the elements. For this reason the oxygen bond energy was determined experimentally for each oxide. The values of q_0 experimentally obtained were then compared with the catalytic activity of the oxides with respect to the reaction of homomolecular isotopic exchange of oxygen ($^{16}O_2 + {}^{18}O_2 \rightarrow 2{}^{16}O^{18}O$) and also with respect to various oxidation reactions involving molecular oxygen.

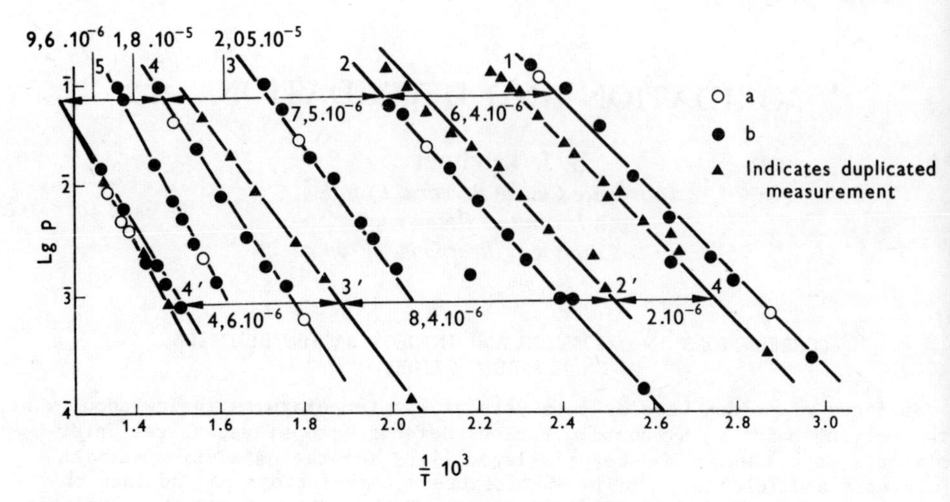

Fig. 1. Oxygen pressure as a function of temperature for NiO: (1-6) first experiment with 16.1 g of oxide; (1'-4') second experiment with 8.67 g of oxide; (a) measurements at reduced temperature; (b) measurement at higher temperature. (The numbers between the curves correspond to the quantity of oxygen (mol) removed from the specimen surface). After SAZONOV, B.A., POPOVSKII, V.V. and BORESKOV, G.K. (1968). *Kinetika i Kataliz*, 9, 312.

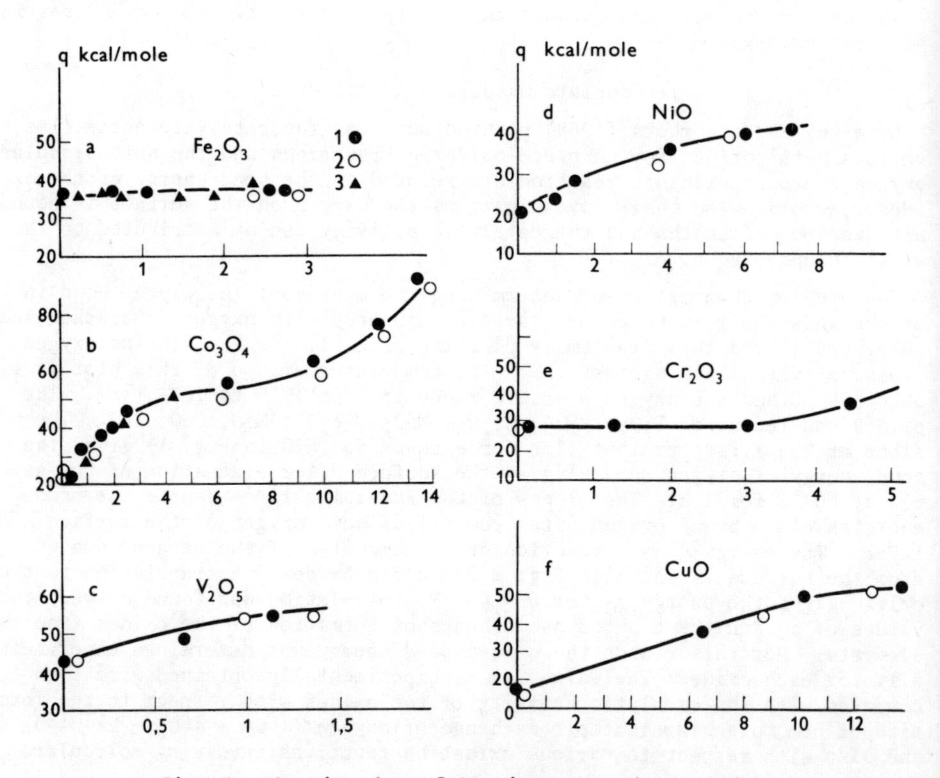

Fig. 2. Continued on following page with caption.

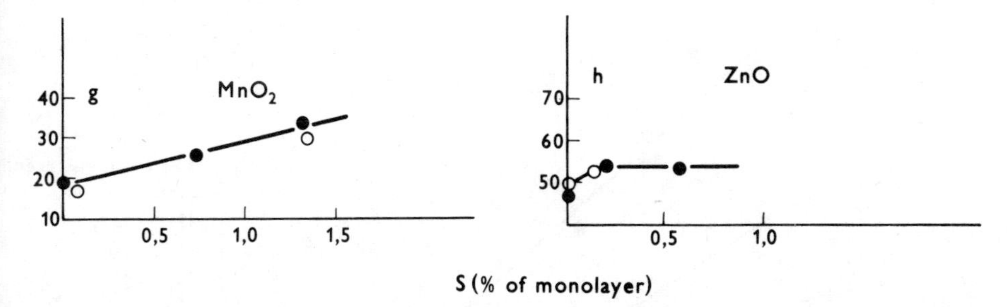

Fig. 2. Bond energy of surface oxygen of various metal oxides as a function of the quantity of oxygen removed: (1) series I measurements —•—; (2) series II measurements —o—; (3) series III measurements —▲— (on Figs. Figs 2c and 2d where the open circles refer to the series III measurements). After SAZONOV, B.A., POPOVSKII, V.V. and BORESKOV, G.K. (1969). *Kinetika i Kataliz*, 9, 312.

Fig. 3 shows the activation energy of homomolecular oxygen isotopic exchange reaction (curve I), of methane oxidation (curve II) and of hydrogen oxidation (curve III) as a function of q_0. In all cases the energy of activation of the reaction is directly proportional to the increase in the oxygen bond energy q_0. The positive sign of the slopes is an indication that

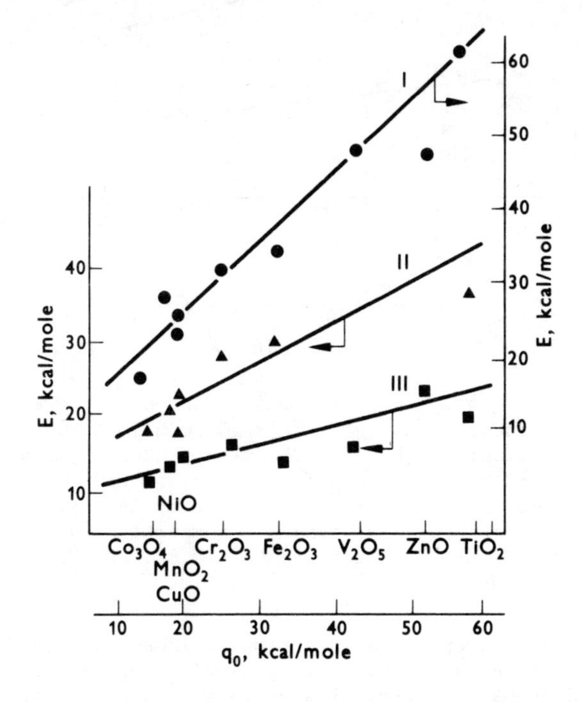

Fig. 3. Activation energy of a reaction as a function of the surface oxygen bond energy in the case of oxides: (I) isotopic exchange of oxygen; (II) oxidation of methane; (III) oxidation of hydrogen. After SAZONOV, B.A., POPOVSKII, V.V. and BORESKOV, G.K. (1968). *Kinetika i Kataliz*, 9, 312.

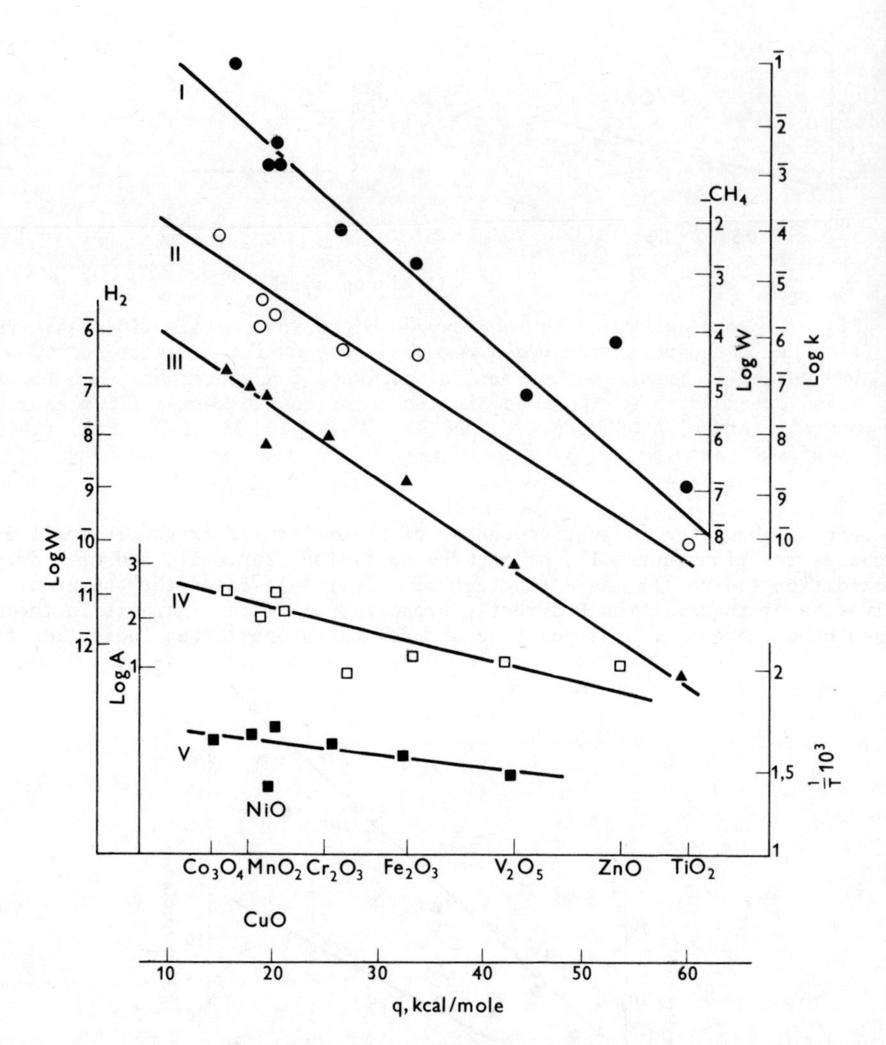

Fig. 4. Catalytic activity of the oxides and bond energy of surface oxygen: isotope exchange of oxygen $\log[k(\mathrm{gO_2/m^2h})]$ at 300 °C, $p\mathrm{O_2}$ = 40 torr; (II) oxidation of methane $W_{\mathrm{CH_4}}$ (l $\mathrm{CH_4/m^2h}$) at 300 °C, $C_{\mathrm{CH_4}}$ = 1 vol. %; (III) oxidation of hydrogen, $W_{\mathrm{H_2}}$(mol $\mathrm{H_2/m^2h}$) at 300 °C, $C_{\mathrm{H_2}}$ = 0.2 vol. %; (IV) oxidation of carbon monoxide A (arbitrary units); (V) oxidation of propylene $10^3/T$ (reciprocal of absolute temperature at which reaction rate is 1.5×10^{-6} mol $\mathrm{O_2/m^2s}$). After SAZONOV, B.A., POPOVSKII, V.V. and BORESKOV, G.K. (1968). *Kinetika i Kataliz*, 9, 312.

the rate-determining step of the reaction includes the rupture of the oxygen bound to the catalyst. The different values of the slope result from the different degrees of deformation of this bond in the activated complex.

A similar rectilinear relation holds when the logarithm of the rate constant (per m^2) for the same reactions (and also, in addition, for CO and propylene oxidations) is plotted against the oxygen bond energy, as shown in Fig. 4, where curve I is for isotopic O_2 exchange, curve II for oxidation of CH_4, curve III for oxidation of H_2, curve IV for oxidation of CO and curve V for oxidation of propylene.

A simple correlation between the catalytic activity and oxygen bond energy cannot be generally universal. The activation energy of the rate-determining step of the oxidation reaction may depend not only on the energy of the oxygen bond with the catalyst but also on the bond energy of other partners of the reaction. The variation of this last bond energy in a series of oxides may not agree with the oxygen bond energy and may therefore lead to a significant distortion of activities in a given sequence of oxide catalysts.

It seems also that it is more difficult to establish the correlation between the catalytic properties and the oxygen bond energy when the selectivity for the desired products is considered. For instance (Boreskov and coworkers, 1968), the specific activity of the oxides of Group IV of the periodic table in the total oxidation of methanol agrees with the energy of the surface oxygen bond. However, the selectivity of this oxidation into formaldehyde sharply declines with rising bond energy of the oxygen with the catalyst. It seems therefore that in order to obtain the products of partial oxidation the reactivity of oxygen on the surface must be sufficiently high to guarantee the required oxidation rate of the initial reagent, but not so high as to oxidize the valuable products of partial oxidation.

A new correlation concerning the energy of the oxygen bond and the activity of various oxides for the overall oxidation of toluene comes from the work of Germain and coworkers (1972) interpreted by Vijh (1972). Germain *et al.* studied the oxidation of toluene in air at 400-450 °C on 19 metallic oxides. The catalysts were classified according to the rate of the overall oxidation of toluene in the sequence shown in Table I. The partial oxidation

TABLE I†

Overall Rate of Oxidation of Toluene in Air at 400 °*C*

$Co_3O_4 > CuO > Mn_2O_3 > Fe_2O_3 > Cr_2O_3 > U_3O_8 > NiO > V_2O_5 > TiO_2 > ThO_2$
$> WO_3 > ZnO > SnO_2 > MoO_3 > Bi_2O_3 > Ta_2O_5 > Nb_2O_5 > ZrO_2 > Sb_2O_4$

Initial Selectivity in Products of Partial Oxidation

(benzaldehyde > 50% for V, Mo and W oxides)

$V_2O_5 > MoO_3 > WO_3 > U_3O_8 > Sb_2O_4 > Cr_2O_3 > NiO, TiO > SnO_2, Nb_2O_5$
$> ZnO, Mn_2O_3 > CuO > Ta_2O_5 > Co_3O_4, ThO_2, ZrO_2 > Fe_2O_3, Bi_2O_3$

product was mainly benzaldehyde (more than 50% on the oxides of V, Mo, W) and also some acids such as benzoic, phthalic and maleic. The inital selectivity into products of partial oxidation gives a different sequence, also shown in Table I. If the logarithm of the rate of the overall oxidation of toluene is plotted as a function of the heat of formation of the oxides used as catalysts, divided by the number of oxygen atoms in the oxide (Ozaki and coworkers, 1966), no definite correlation is found (Fig. 5). On the contrary,

† From GERMAIN, J.E. and LAUGIER, R. (1972). *Bull. Soc. Chim. France*, p. 54.

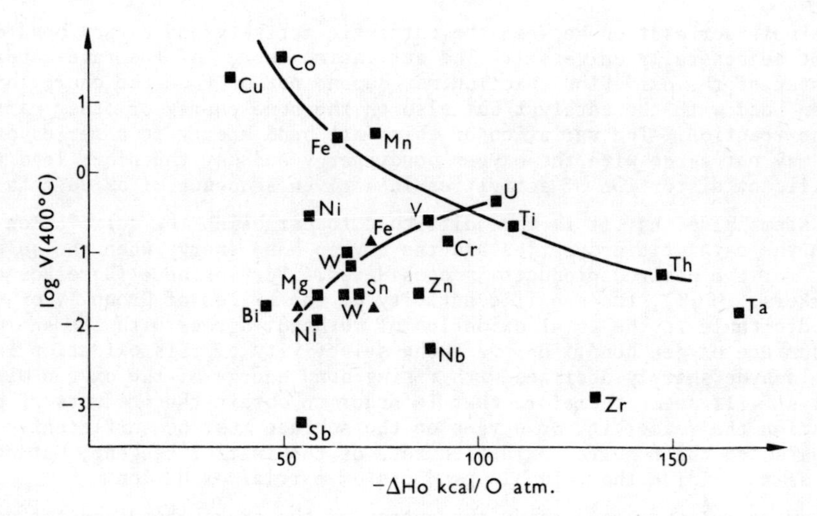

Fig. 5. Attempted correlation between catalytic activity for toluene oxidation and the oxygen bond energy. After GERMAIN, J.E. and LAUGIER, R. (1972). *Bull. Soc. Chim. France*, p. 54.

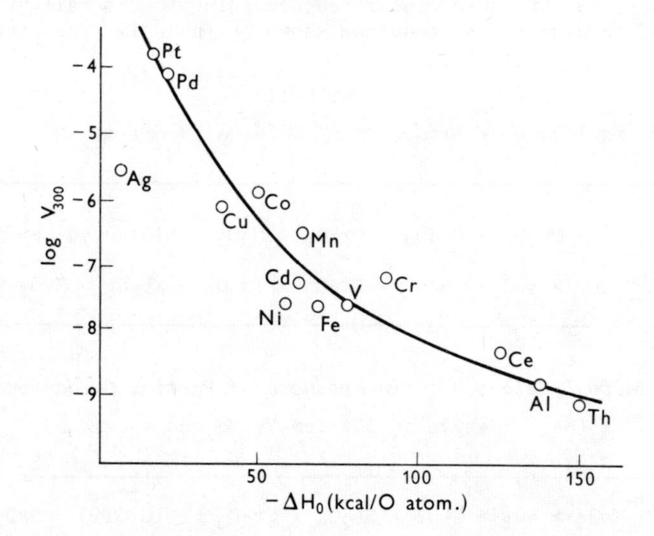

Fig. 6. Correlation between the catalytic activity and the heat of catalyst formation (ΔH_0)(L series). After MOROOKA, Y. and OZAKI, A. (1966). *J. Catalysis*, 5, 116.

for the complete oxidation of propylene, Ozaki *et al.* found a good correlation based on the same parameter (Fig. 6). The larger this binding energy, the less active is the catalyst. Such a correlation is related to the mobility of the oxygen in the lattice of the oxide, but it concerns complete oxidation. Also Germain and coworkers (1971) assume that one of the criteria of selectivity in the partial oxidation of C_3 and C_4 olefins must be related to the stability of π-allyl complexes adsorbed onto a surface of the catalyst. A weak bond between adsorbed olefin and the catalyst is favorable to the partial oxidation whereas a strong bond results in complete oxidation.

However, in the oxidation of toluene by Germain, a different picture from that of Fig. 5 is found if the average bond energy of a metal oxide is calculated by the equation proposed by Vijh and Lenfant (1971):

$$b(\mathrm{MO}) = \frac{-\Delta H_m + \Delta H_S + \frac{n}{4}\Delta H_D}{n}$$

where b(MO) is the average bond energy, ΔH_m is the heat of formation per mole (standard state) of the metal oxide, ΔH_S is the heat of sublimation per mole of the metal, and ΔH_D is the heat of dissociation of the oxygen molecule (in the gas phase) to give two oxygen atoms. Finally n is the number of charges transferred from metal to the oxygen atoms in forming one molecule of the oxide (e.g. n is 2 for NiO, 4 for ZrO_2, 10 for Ta_2O_5, etc.). The b(MO) value thus calculated is then the heat of atomization per equivalent of the metal oxide and represents the enthalpy change, per bond, in the reaction:

$$\mathrm{MO_{(s)}} \longrightarrow \mathrm{M_{(g)}} + \mathrm{O_{(g)}}$$

Fig. 7 shows the plot of b(MO) values for oxides used by Germain against the logarithm of the rate of the overall oxidation of toluene (at 450 °C). A

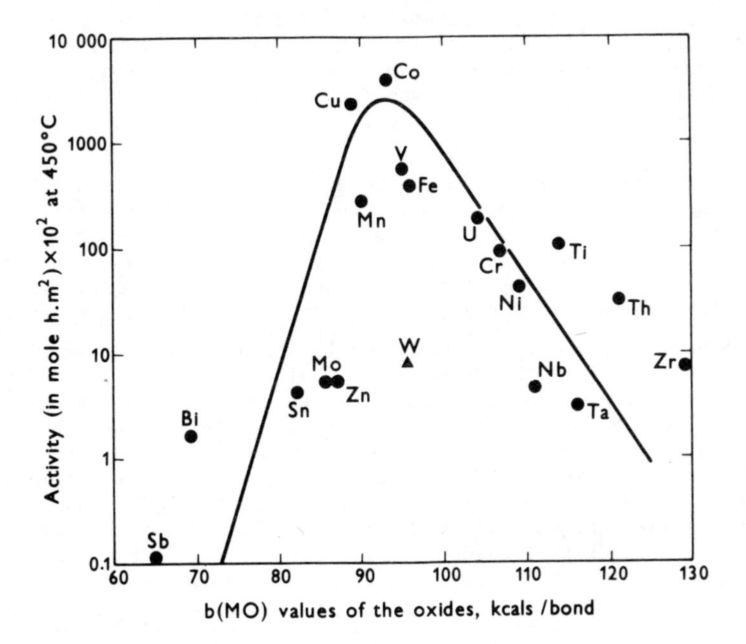

Fig. 7. Dependence of catalytic activity on oxide energy parameter b(MO). After VIJH, A.R. (1972). *J. Chim. Phys.*, 69, 1695.

volcano-type correlation is found (Vijh, 1972), quite well known in heterogeneous catalysis. For the oxides of Sb, Bi, Sn, Mo, Zn, Mn, and Cu (to be called class I), the activity increases with increasing b(MO) values, whereas the oxides of Co, V, Fe, U, Cr, Ni, Ti, Nb, Th, Ta, and Zr (to be called class II here) show the opposite trend.

This would suggest that for the class I oxides the rate-determining step in the overall oxidation is perhaps the formation of an MO bond (between an M site of the catalyst and oxygen from the gas phase), whereas class II oxides probably involve the rupture of an MO bond as the rate-determining step. This behavior could also be interpreted according to the view of Bond (196 as follows: the rate on oxides of class I (Sb, Bi, Sn, Mo, and Zn) is low because the coverage by adsorbed oxygen is low. The rate on oxides of class II (Ta, Zr, Nb, Th, Ni, and Ti) is low because the adsorbed oxygen is too strongly bound. For optimally high rates, at the top of the curve of Fig. 7 (i.e. the oxides of Mn, Cu, Co, V, Fe, U, and Cr) a reasonable coverage by adsorbed oxygen as well as a moderately high value of b(MO) is required. These findings may provide a conceptual guideline for interpreting the oxidation activity of various oxides.

The Deoxidation Reactions

The reactions of deoxidation such as the decomposition of N_2O or of NO on oxides were first considered on the basis of semiconductor properties of oxide catalysts. The division was into good catalysts (semiconductors of p-type), and bad ones (semiconductors of n-type), with insulators mainly occupying an intermediate position. Instead of this crude division, Winter (1969) was able to show that the catalytic activity is controlled mainly by the oxide type and the lattic parameter. In a previous work Winter (1968) demonstrated that the heteronuclear exchange between $^{18}O_2$ in the gas phase and the surface of *stable* metallic oxides is controlled by the desorption step. Therefore a critical parameter in the activated state must be the O-O distance which should be determined by, or at least strongly influenced by, the nearest O-O distance in the oxide surface. Indeed, the activation energy of the exchange reaction was found to fall in a rectilinear manner with increasing size of the unit cell in oxides of the same crystal structure.

For the decomposition of N_2O the desorption of oxygen is not the rate-determining step. The slow step is the decomposition of adsorbed N_2O. However, the overall rate is determined, in part, by the number of reaction sites available for N_2O chemisorption; this number is determined by competition for the surface sites between O_2 and N_2O in their equilibrium adsorptions. In other words oxygen acts as a poison for N_2O decomposition.

Since the oxygen desorption reaction is dependent on the lattice parameter of the oxide catalyst, the decomposition of N_2O will be also. This is shown in Fig. 8 concerning the activation energies of decomposition of N_2O on 15 oxides of the type M_2O_3; a rectilinear decrease in the activation energy with lattice parameter expressed as (molecular volume)$^{1/3}$ is observed. This work was extended by Winter (1970) to other oxides. The corresponding plot (Fig. 9) clearly divides the oxides into three types: MO, M_2O_3, and MO_2. The same correlation is obtained if instead of activation energy the rate of decomposition of N_2O ($\log K$) is considered. It is to be noted that the reaction parameters (E, $\log K$) which are correlated here to lattice parameter and in particular to the minimum O-O distance could be equally correlated to any chemical or physical property which changes regularly with lattice parameter.

The desorption of an O_2 molecule may occur only from pairs of anion sites at the minimum O-O distance. In most cases the reaction has required the

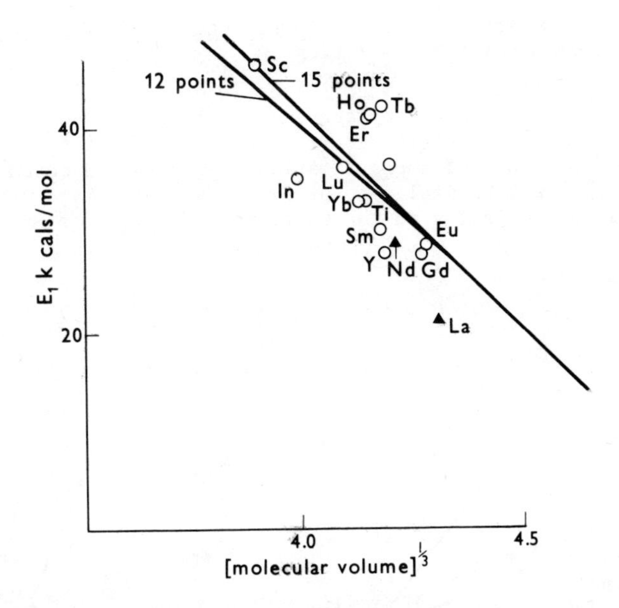

Fig. 8. Dependence of the activation energy of N_2O decomposition on the crystal lattice parameter. The crystal structures of oxides of La and Nd, ▲, differ from the structure of other oxides. After WINTER, E.R.S. (1969). *J. Catalysis*, 15, 144.

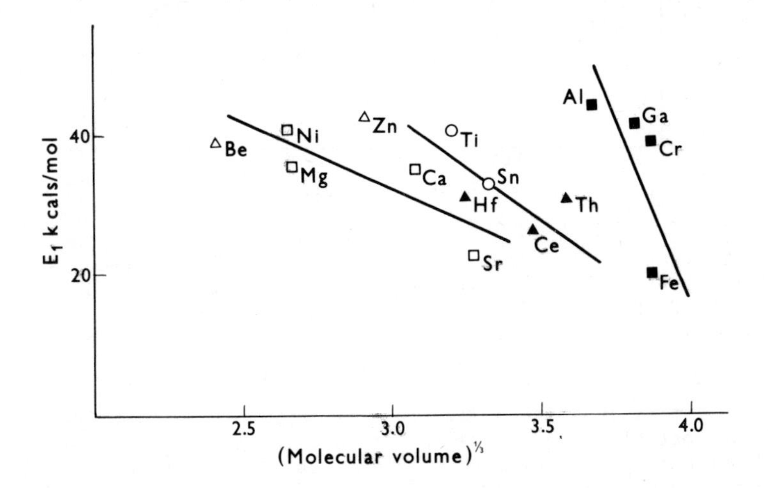

Fig. 9. Variation of the activation energy with lattic parameter and crystal type for the N_2O decomposition. o, □, Δ, ■, ▲ represent different oxide structures.

use of rather high temperatures (which are a substantial fraction of the melting points of the oxides). It is unlikely that, at such temperatures, crystal field effects will be of much importance in determining the course or energetics of the reaction, since appreciable surface mobility must be present under reaction conditions. Such mobility must involve the compensation, or neutralization, of lattice forces much greater in magnitude than the crystal field stabilization forces.

Though the rate-determining step in the NO decomposition (Winter (1971)) is the adsorption of NO on a surface double site, the desorption of oxygen is still the step which interferes with the main reaction by its poisoning effect. The dependence of the activation energy for NO decomposition upon lattice parameter and crystal type of oxide catalysts is shown in Fig. 10. There is a close similarity to the corresponding plot for N_2O decomposition.

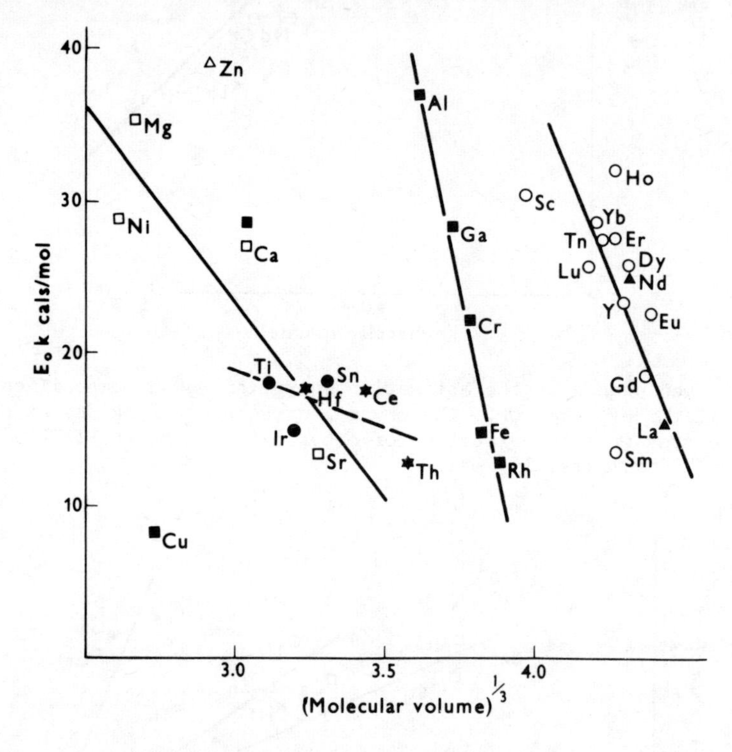

Fig. 10. Dependence of E_0 for NO decomposition upon lattice parameter and crystal type. o, □, Δ, ▲, ●, ✱ represent different crystal structures. After WINTER, E.R.S. (1971). *J. Catalysis*, 22, 158.

Partial Oxidation Reactions

Acrolein synthesis by the oxidation of propylene has been investigated as one of the most basic reactions in a mild oxidation. The studies have been carried out mainly over bismuth molybdate catalysts, which are industrially important. Adams and Jenkins (1963) offered a mechanism in which the oxidation of propylene into acrolein proceeds via a π-allyl intermediate. Schuit and coworkers (1966) suggested that the olefin can be bound at the anion vacancy adjacent to Mo ions. Sachtler and de Boer (1964) have suggested a reducibility parameter in order to interpret the activities and selectivities of mixed-oxide catalysts. This reducibility is characterized by the temperature at which the oxides begin to be reduced by hydrogen. However, no clear correlations were established for the selectivity in the partial oxidation. In general, the catalytic activity of these various oxides appears to increase with decreasing energy characterizing the removal of oxygen from the surface, whereas an inverse tendency is found for the selectivity.

The situation seems to be better for the selective decomposition of 2-propanol into acetone over manganese oxides, a type of redox reaction. A relation between the oxygen content of the manganese oxide catalyst and their activity was established by McCaffrey and coworkers (1972). Fig. 11 plots the logarithm of the rate of dehydrogenation of 2-propanol either at 175 °C or at 250 °C v. the O/Mn ratio of the oxides examined. The reciprocal of the binding energy of the Mn-O bond is included also as an ordinate in this Figure. This binding energy is defined as the standard enthalpy ΔH^{0}_{295} of the reaction:

$$\frac{1}{m} Mn_xO_y + \frac{1}{2} O_2 \longrightarrow \frac{1}{m} Mn_xO_{y+m}$$

where the higher oxide represents either bulk or surface compound (Klier, 1967). The Figure shows a correlation between the logarithm of the rate of dehydrogenation and the reciprocal of the binding energy of the last gram

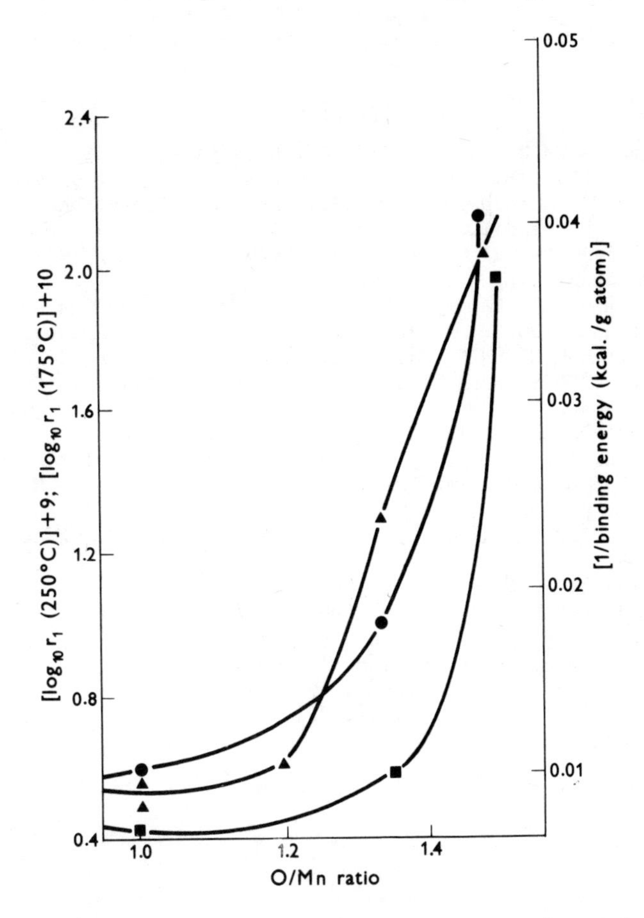

Fig. 11. Log_{10}(rate of dehydrogenation of 2-propanol) and (binding energy)$^{-1}$ v. O/Mn ratio. (●) (binding energy)$^{-1}$; (▲) rate of dehydrogenation at 250 °C; (■) rate of dehydrogenation at 175 °C. After McCAFFREY, E.F., KLISSURSKI, D.G. and ROSS, R.A. (1972). *J. Catalysis*, 26, 380.

atom of oxygen added to the metal oxide. Earlier Klissurski (1968) established a similar relation between the selectivity for oxidation of methanol to formaldehyde on transition-metal oxides and similar energy values. These results are therefore of the same type as those of Boreskov and coworkers (1968, 1969) for the *complete* oxidation of methane and propylene on transition metal oxides, previously examined.

In conclusion, it seems that for a better correlation between the catalyti activity in a complete oxidation, or even in a partial oxidation, the parameter which needs to be defined more precisely is the binding energy of the oxygen (at the surface lattice of the oxide) which is actually reacting in the oxidation reaction. Although the experimental results of Boreskov for the binding energy of oxygen on oxides leave little doubt as to the principl the values calculated from the thermochemical data, in the absence of the experimental values, need to be revised in order to take into account the binding energy of the *reactive* oxygen.

Finally, a second problem arises when the selectivity between two or more products in oxidation is considered. The bond energy of the reaction products (or of some stable intermediates) to the solid may also be needed.

REFERENCES

ADAMS, C.R. and JENKINS, T.J. (1963). *J. Catalysis*, 2, 63.
BATIST, Ph., LIPPENS, B.C. and SCHUIT, G.C.A. (1966). *J. Catalysis*, 5, 55.
BOND, G.C. (1962). *Catalysis by Metals*, (Academic Press, New York).
BORESKOV, G.K., POPOVSKII, V.V. and SAZONOV, B.A. (1971). *Proceedings of the 4th International Congress on Catalysis, Moscow, 1968*, (Akakemiai Kiado, Budapest), p. 439, (paper 33).
GERMAIN, J.E. and LAUGIER, R. (1971). *Bull. Soc. Chim. France*, p. 650.
GERMAIN, J.E. and LAUGIER, R. (1972). *Bull. Soc. Chim. France*, p. 541.
GRAVELLE, P.C. and TEICHNER, S.J. (1969). *Adv. Catalysis*, 20, 167.
McCAFFREY, R.F., KLISSURSKI, D.G. and ROSS, R.A. (1972). *J. Catalysis*, 26, 380.
MOROOKA, Y. and OZAKI, A. (1966). *J. Catalysis*, 5, 116.
SACHTLER, W.M.H. and DE BOER, J.H. (1965). *Proceedings of the 3rd International Congress on Catalysis, Amsterdam, 1964*, (North Holland, Amsterdam), p. 252.
SAZONOV, B.A., POPOVSKII, V.V. and BORESKOV, G.K. (1968). *Kinetika i Kataliz*, 9, 312.
VIJH, A.R. (1972). *J. Chim. Phys.*, 69, 1695.
VIJH, A.R. and LENFANT, P. (1971). *Can. J. Chem.*, 5, 809.
WINTER, E.R.S. (1968). *J. Chem. Soc. A.*, p. 2889.
WINTER, E.R.S. (1969). *J. Catalysis*, 15, 144.
WINTER, E.R.S. (1970). *J. Catalysis*, 19, 32.
WINTER, E.R.S. (1971). *J. Catalysis*, 22, 158.

REACTION SCHEMES AND COORDINATES

W. Keith Hall
Gulf Research and Development Company, P.O. Drawer 2038, Pittsburgh, Pennsylvania 15230, USA

A purpose of the present conference is to develop lines of communication between workers in the three fields of heterogeneous, homogeneous, and enzyme catalysis. Another is to seek ways in which workers in the three fields can draw upon the work of each other to establish common chemistry. Another concern must be the ways in which the intervention of a solid surface alters the resulting chemistry. Finally, we must seek out the kinds of work which need to be done to accomplish these objectives. The purpose of the present brief communication is to suggest two approaches to these problems.

The salient attractive features of enzyme catalysis are the extremely high activities achieved and the great selectivities to effect one desired reaction to the exclusion of others. The chief limitation to its wider applications is the fact that enzymes are effective in aqueous media only, and over a very narrow temperature range. The great potential for industrial applications of homogeneous catalysis lies in the hope that systems can be devised which will simulate the attractive features of enzyme catalysis, but without these limitations. Thus, a consideration of the chemical principles which may lead to selectivity becomes pertinent. These include not only effects of site composition and geometry, but energetic factors as well. Little attention has been paid to the latter in spite of the fact that they frequently control the selectivity.

Consider the case of a single reactant forming an intermediate species which can decompose into either of two product molecules. When a true intermediate exists it is in a metastable state which resides in a potential well near the top of the reaction coordinate. To escape this well, it must pass over one of three barriers — back to the reactant or to one of the products. Selectivity may be controlled by the relative heights of the three barriers.

Carbonium-ion chemistry is one of the few cases where an entire reaction scheme has been carried over from homogeneous systems into heterogeneous catalysis. Knowing the scheme, the reaction coordinates may be constructed with the help of experimental data for the isomerization of n-butenes (Fig. 1). Note that both the stability of the intermediate and the selectivity for product formation vary with the strength of the interaction between substrate and acid. In principle, this sort of a picture can be developed for any reaction involving a common intermediate and such semiquantitative data afford a means of comparing vastly different catalytic systems. Moreover, if a reaction does not invove a single common intermediate, the reaction coordinates derived from the experimental data usually will be inconsistent with this simple picture. Such cases have been found in the writer's laboratory for butene isomerization, both for alumina catalysts and for the homogeneous tris-triphenylphosphine-rhodium chloride-ethanol-HCl system. The point to be made, however, is that by carefully choosing a few reactants which can produce several products, and by studying the selectivity as a function of temperature, a formalism can be developed by which common chemistry can be established for different catalyst systems.

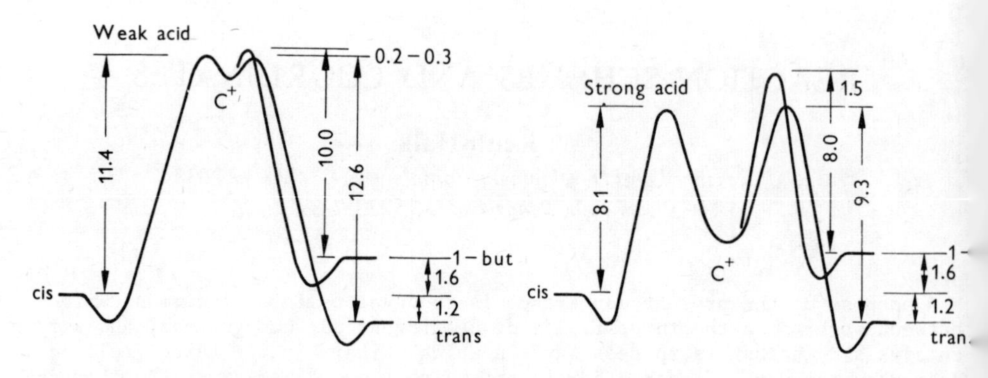

Fig. 1. Reaction coordinates for butene isomerization over heterogeneous acid catalysts. The numbers refer to kcal/mol.

The situation becomes more complicated than that described above when more than one unstable intermediate exists along the reaction coordinate, e.g. several different carbonium ions. Differences in statistical (geometric) factors may also be expected when homogeneous systems are compared with heterogeneous ones. For example, temperature-independent *cis*/*trans* ratios near unity are always found for 1-butene isomerization over solids when the reaction follows a carbonium-ion mechanism. This is because of the steric requirements imposed by the solid which make the formation of these compounds equally probable. This need not be the case when the catalyst is in solution, although it could occur under special circumstances.

The above discussion was rendered straightforward because the reaction scheme is well known for carbonium-ion chemistry. Frequently this is not the case with other catalytic systems, or else several different mechanisms have been proposed. Thus, supplemental information concerning the course of the reaction is required before a meaningful reaction coordinate can be drawn. A reaction scheme, if not its mechanism, may be deduced from experiments involving the judicious use of tracers. Unfortunately, a general prescription cannot be written for this purpose; many workers have devised experiments which have proved useful with particular systems. What is needed to meet the present objectives is to search for comparable chemistry for the same reactions in all three types of catalysis. By way of illustration, there follows a brief summary of some reaction schemes, derived from tracer experiments, for the formation of partial oxidation products in the oxidation of simple olefins over noble metal catalysts. Also, some comments are made concerning the relations of these results with those from other catalytic systems. It should be recognized that all of these processes take place to some extent over most of the metals mentioned. This is a case, then, where improved selectivities potentially could be developed by homogeneous catalysts.

In the oxidation of ethylene over Pd and Ir, acetaldehyde and acetic acid are formed with an intramolecular hydrogen shift to form the methyl group, i.e.

$$50\%\ C_2H_4 \xrightarrow{O} CH_3CHO \xrightarrow{O} CH_3COOH\ (D)$$

$$+$$

$$50\%\ C_2D_4 \xrightarrow{O} CD_3CDO \xrightarrow{O} CD_3COOD\ (H)$$

$$\downarrow\uparrow$$

(minimal intermolecular exchange)

The acetaldehyde does *not* result from isomerization of ethylene oxide. The latter, labelled with ^{14}C, passes through the reactor during reaction of C_2H_4 without decomposing or entering into any of the products. A similar 1-2 intramolecular hydrogen shift is known to occur in the Wacker process ($PdCl_2$). No $^{16}O^{18}O$ is observed when mixtures of $^{16}O_2$ and $^{18}O_2$ are used in the reaction, although the CO_2 is equilibrated. The same activation energy is found for total oxidation as for acetic acid formation, yet tracers show that most of the CO_2 is produced by a parallel pathway and not by oxidation of $CH_3{}^{14}CHO$.

Relatively small amounts of propionic acid and acetone are formed from propylene. Much larger amounts of acetic acid are produced over Ir. This has been traced to:

$$CH_3CH{=}{}^{14}CH_2 \xrightarrow{O} CH_3COOH + {}^{14}CO_2$$

Since the 2-butenes yielded about twice as much acetic acid, it is thought that HCHO is a primary product with terminal olefins, and that it is not observed because it reacts to CO_2 and H_2O at a rate many times faster than that for C_3H_6. A related attack at the double bond is known to occur with OsO_4 in aqueous solution to form *cis*-diols.

Selective oxidation to epoxides (C_2H_4 and C_3H_6) is unique to Ag catalysts. Epoxides are also formed in the homogeneous gas-phase reaction of olefins with ground state (triplet) oxygen atoms. In both the homogeneous and Ag-catalysed reaction the double bond is opened in the transition state so that *cis-trans* isomerized epoxides are formed, i.e.:

$$(D)(H)C{=}C(D)(H) \xrightarrow{O} (D)(H)C\!-\!C(D)(H)\ [\text{epoxide, } cis] + (D)(H)C\!-\!C(H)(D)\ [\text{epoxide, } trans]$$

However, in the gas-phase reaction a nearly equal amount of the corresponding carbonyl isomer (in this case acetaldehyde) is produced whereas this is not observed in the Ag-catalysed products. Thus, tracers have revealed this interesting state of affairs: both products are formed in the homogeneous reattion and the one product over Ag and the other over Pd.

Attack at a paraffinic hydrogen leads to the formation of acrolein from C_3H_6 (and probably butadiene from 1-butene) via a symmetric intermediate. The data for Rh catalysts may be represented by

$$CH_3CHCD_2 \xrightarrow{yk_H} \dot{C}H_2CHCD_2 \xrightarrow{f} CH_2CHCDO$$
$$\dot{C}H_2CHCD_2 \xrightarrow{(1-f)} CD_2CHCHO + H_2O$$
$$CH_3CHCD_2 \xrightarrow{(1-y)k_H} HO_2CH_2CHCD_2 \xrightarrow{1} CD_2CHCHO + H_2O$$

and

$$\begin{array}{l} CD_3CHCH_2 \xrightarrow{xk_D} \overline{CD_2\dot{C}HCH_2} \xrightarrow{(1-f)} CD_2CHCHO \\ \overline{CD_2\dot{C}HCH_2} \xrightarrow{f} CH_2CHCDO + D_2O \\ CD_3CHCH_2 \xrightarrow{(1-x)k_D} DO_2CD_2CHCH_2 \xrightarrow{1} CH_2CHCDO + D_2O \end{array}$$

where $k_H/k_D \approx 2.5$, $x = 0.79 \pm 0.05$, $y = 0.83 \pm 0.09$ and $f = 0.49 \pm 0.04$. Similar results are obtained over Au catalysts except that now $x = y = 1.0$. The same scheme may be generated from results of experiments with $CH_3CH{=}^{13}CH_2$. The large primary isotope effect for removal of an H or D from the methyl group shows that breaking of this bond is important in the rate-determining step. A similar result has been obtained for oxide catalysts by Sachtler and by Voge and Adams. These workers also found a primary isotope effect for the removal of H or D from the symmetric intermediate which appeared to be absent over the noble-metal catalysts, i.e. $f \approx 0.5$.

Finally, the reaction of acrolein to $CO_2 + H_2O$ is facile over the noble-metal catalysts. Although as yet unproved, it is suspected that this is due to the attack of the double bond in a manner similar to that found for Ir. Thus, the distribution of partial oxidation products found for noble metals can be interpreted in terms of the chemistry outlined above. Unknown, are the reasons why the different metals favor one or another of these reactions. If homogeneous counterparts of these schemes can be found, it may prove the key to further progress in this area. Even in the case where the same products are found to be formed by a different scheme, the new knowledge should prove helpful in solving the puzzles which remain in the heterogeneous system.

PART III
HOMOGENEOUS CATALYSIS

HOMOGENEOUS CATALYSIS BY METAL IONS AND COMPLEXES

Joseph Chatt
School of Molecular Sciences, The University of Sussex, Falmer, Brighton, BN1 9QJ, England

and

Jack Halpern
Department of Chemistry, The University of Chicago, Chicago, Illinois 60637, USA

METAL COMPLEXES AND THE NATURE OF ACTIVE SITES

Metal ions exist in solution only as complexes with a full complement of attached groups known as ligands. In aqueous solution water can be the ligand molecule, e.g. $[Ni(H_2O)_6{}^{2+}]$. The reacting substrates must therefore be potential ligands or dissociate to produce ligands which enter the coordination shell of the metal, either additional to the normal complement of ligands or by the displacement of some of the ligand molecules already there. The substrates are usually nucleophilic substances, molecules or anions, generally with π-electronic systems or electron pairs on a reasonably accessible energy level.

The attachment of a molecule to a metal ion does not necessarily render it more reactive chemically. Because of the positive charge on the metal ion, a ligand is often activated for nucleophilic, but deactivated for electrophilic attack. Thus, very early in the study of the acid-catalysed hydration of olefins it was noted that the rate of hydration was proportional to the concentration of free olefin in the solution and that the addition of silver ion, which lowered the concentration of free olefin by complexing part of it, correspondingly lowered the rate of hydration. Similarly diimine, NH=NH, which is a highly reactive and unstable molecule in the free state forms a very stable complex, $[WBr_2(N_2H_2)(Ph_2PCH_2CH_2PPh_2)_2]$, in which form it resists reduction to hydrazine or ammonia. Nevertheless, in the above complex it can be acylated by organic acid chlorides.

Thus, commonly the function of the metal ion catalyst is not to activate the substrate molecule through coordination but rather to stabilize highly reactive species such as H^- or CH_3^- and thereby make them accessible reaction intermediates. Metal complexes which are active catalysts are commonly those whose ligands are somewhat labile and readily displaced by substrate molecules. Alternatively, catalytic complexes whose central atoms have fewer electrons than the closed shell of 8 or 18 electrons (for example 16 valence electron square planar d^8 complexes such as $[Rh^{I}Cl(PPh_3)_3]$), although they are stable, are capable of taking up additional ligand molecules.

Metal ions show well recognized affinities for certain ligand atoms and may be classified accordingly. Some metals tend to attach themselves most strongly to the ligand atoms nitrogen, oxygen or fluorine, as compared with phosphorus, sulfur or chlorine, whereas others prefer the elements phosphorus, sulfur and chlorine. The first type has been labelled Class (a) ions or 'hard' acids, and the second Class (b) ions or 'soft' acids. The ligands have also been named hard or soft bases, according as to whether their affinities are with the Class (a) or Class (b) metal ions. Thus, metal ions will tend to bring together, through coordination, ligands from the same Class, the Class (a) metal ions tending to bring in hard ligands which are often no very reactive, and the Class (b) to bring in soft ligands, which include unsaturated hydrocarbons, carbon monoxide, hydride ion, the oxygen molecule an such reactive substances. Thus, a very extensive area of homogeneous catalysis involves the reactions of unsaturated compounds, carbon monoxide, hydrogen and such species on Class (b) metal ion complexes. Most transition metals in their lowest oxidation states have Class (b) character and a great number of olefin reactions occur on such transition-metal complexes.

In general it is found that the greatest variety of catalysed reactions of olefins such as isomerization, hydrogenation, hydroformylation, hydrosilation and the formation of low polymers occurs especially on complexes of the heavier metals of Group VIII. Compounds of the lighter early transition metals, especially titanium and vanadium treated with very reactive organometallic reagents, e.g., $AlEt_3$, give homogeneous and heterogeneous catalysts (Ziegler-Natta catalysts) for the production of high polymers. Under similar conditions some intermediate metal compounds, especially the halides of tungsten, molybdenum and rhenium, give homogeneous catalysts for olefin metathesis. The mechanism of this reaction is probably the least understood of all the widely studied reactions of olefins catalysed homogeneously by transition-metal compounds.

An important aspect of transition metal ion catalysis is the effect which one ligand in a complex can exert upon another, especially upon the one in the *trans* position to itself. Some ligands have the property of rendering the *trans* ligand very labile (*trans*-effect). Thus, types of complex which are normally inert can be made reactive by introduction of ligands of high *trans*-effect. Especially prominent amongst such ligands are hydrocarbon radicals, carbon monoxide, and hydride ion (which are also often substrates in many important reactions); other well-recognized ligands for this purpose are tertiary organic phosphines, and $SnCl_3^-$ which is obtained by the reaction of $SnCl_2$ on a chloro-complex, e.g. $Pt\text{-}Cl + SnCl_2 \rightarrow PtSnCl_3$. Thus, halide complexes, when they are poor catalysts, might be improved considerably by a reaction with hydrogen, sodium borohydride or $SnCl_2$.

The oxidation state of the metal has a marked influence on the lability of a complex. Thus, octahedral d^3- (and low-spin d^6-) complexes are often inert e.g. $[Cr(H_2O)_6]^{3+}$, but a change of oxidation state by one unit produces $[Cr(H_2O)_n]^{2+}$ which is a labile complex. One of the oldest examples of catalysis by changing the oxidation state of the ion is found in the dissolution of anhydrous $CrCl_3$ in water. This chemical process is extraordinarily slow, but in the presence of a reducing agent which produces a trace of Cr^{2+}, dissolution occurs rapidly.

The above-stated general principle that like ligands tend to come together on the same complex has some notable exceptions. Ammonia and water are both hard bases. Thus, one would expect the d^6-octahedral complex $[Ru(NH_3)_5(H_2O)]$ to be very stable except to oxidation, but in fact the water is displaced remarkably readily by certain somewhat inert soft bases, e.g. carbon monoxide and molecular nitrogen, even in aqueous solution. Thus, there are important factors outside the generalities that like ligands tend to come together on appropriate types of metal ions. Similar violations of the other generalities discussed above also are known. This introduces a specificity, often difficult to understand, into a particular metal site. It is often this specificity, depending critically on a unique combination of metal and ligands, which appears to give rise to the most active catalytic sites, and this is a very important factor in catalysis by metalloenzymes.

REACTIVITY PATTERNS RELATED TO HOMOGENEOUS CATALYSIS

Among the various types of reactions which metal ions and complexes undergo, the following deserve mention as being especially relevant to the roles of such ions and complexes in homogeneous catalysis.

Substitution

Catalysis commonly requires coordination of at least one of the reactants to the catalytic metal site, a process which is often achieved through substitution or replacement of ligands in the catalyst complex. Such substitution may occur either by dissociative or associative mechanisms (characteristic of octahedral and square planar complexes, respectively) and must be rapid for the catalytic efficiency to be high. Some of the factors that influence the labilities of complexes (including the widespread and very important *trans*-effect) have already been mentioned.

Oxidation-Reduction

Such reactions are widespread among both the transition and post-transition metals, although it is only the former (in view of their abilities to accommodate different numbers of electrons in their *d*-subshells) that exhibit multiple stable oxidation states differing by one-electron intervals (e.g. Fe^{II}-Fe^{III}). The stable oxidation states of the post-transition metals typically differ by two electrons, e.g. Tl^{I}-Tl^{III}, Pb^{II}-Pb^{IV}, etc..

Low oxidation states are most effectively stabilized by good π-acceptor (or 'soft') ligands such as CO and tertiary phosphines, whereas the highest oxidation states are stabilized by small π-donor (or 'hard') ligands such as O^{2-}, OH^- and F^-. The existence of the following stable chromium complexes illustrates this trend: $[Cr^{0}(CO)_6]$, $[Cr^{III}(H_2O)_6]^{3+}$, $[Cr^{VI}O_4]^{2-}$.

The rates of simple electron-transfer reactions depend upon the nature of the metal ions as well as of the ligands and vary over a very wide range. This is illustrated by the following reactions, for which the second-order rate constants are, $\leqslant 10^{-5}$, ≈ 1, and $> 10^{8}$ $mol^{-1}s^{-1}$, respectively:

$$[Cr(H_2O)_6]^{2+} + [Cr^{*}(H_2O)_6]^{3+} \longrightarrow [Cr(H_2O)_6]^{3+} + [Cr^{*}(H_2O)_6]^{2+} \quad (1)$$

$$[Fe(H_2O)_6]^{2+} + [Fe^{*}(H_2O)_6]^{3+} \longrightarrow [Fe(H_2O)_6]^{3+} + [Fe^{*}(H_2O)_6]^{2+} \quad (2)$$

$$[Fe(phen)_3]^{2+} + [\overset{*}{Fe}(phen)_3]^{3+} \longrightarrow [Fe(phen)_3]^{3+} + [\overset{*}{Fe}(phen)_3]^{2+} \quad (3)$$

'Non-equivalent' redox reactions (e.g. those between 1-electron reductants and 2-electron oxidants) usually occur through stepwise mechanisms involving unstable intermediate oxidation states (e.g. Tl^{II} in the following example), and are accordingly often slow and susceptible to catalysis:

$$\text{Rate-determining: } Fe^{II} + Tl^{III} \longrightarrow Fe^{III} + Tl^{II} \quad (4)$$

$$\text{"Fast": } Fe^{II} + Tl^{II} \longrightarrow Fe^{III} + Tl^{I} \quad (5)$$

$$\text{Overall reaction: } 2Fe^{II} + Tl^{III} \longrightarrow 2Fe^{III} + Tl^{I} \quad (6)$$

Oxidative Addition and Related Reactions

Among the low-spin complexes of transition metals (e.g. carbonyls, cyanides, phophines and organometallic compounds) the closed-shell 6-coordinate d^6 configuration (corresponding to 18 valence electrons) is commonly one of great stability. Accordingly, complexes whose configurations differ from this frequently exhibit high reactivities and readily undergo reactions that restore the stable closed-shell configuration.

Among the important reactions encompassed by the resulting reactivity patterns are those involving the oxidations of certain transition metal complexes by saturated molecules with accompanying incorporation of the fragments of reductive cleavage into the coordination shells, e.g.:

$$2[Co^{II}(CN)_5]^{3-} + CH_3I \longrightarrow [Co^{III}(CH_3)(CN)_5]^{3-} + [Co^{III}I(CN)_5]^{3-} \quad (7)$$

$$[Rh^{I}Cl(PPh_3)_3] + H_2 \longrightarrow [Rh^{III}ClH_2(PPh_3)_3] \quad (8)$$

Such reactions, which are characteristic of 5-coordinate d^7 and 4-coordinate d^8 (as well as certain d^{10}) complexes are commonly designated as *oxidative addition*, reflecting the increase in coordination number that typically accompanies the increase in the oxidation number of the metal. Such oxidative addition reactions are widespread and often provide facile routes for the dissociation of stable molecules such as dihydrogen and organic halides, with accompanying formation of hydrido- and organo-transition metal complexes, respectively. This has obviously important implications for catalysis, for example in catalytic hydrogenation reactions which usually proceed through mechanisms incorporating a step in which the dissociation of H_2 is effected through oxidative addition to the catalyst. The reversals of such oxidative addition reactions (i.e. *reductive eliminations*), which provide effective mechanisms for the formation of new covalent bonds, e.g. H-H, C-H, C-C and C-X (X = Cl, Br, I, S, etc.), also constitute important steps in many of the catalytic processes to be described.

Insertion Reactions

Another class of reactions of transition metal complexes that plays an important role in many catalytic processes is that designated as *insertion*. Such reactions involve the 'insertion' of an unsaturated molecule X (e.g. CO, olefin, acetylene, SO_2, etc.), into a metal-ligand σ-bond M-L (L = H, alkyl, aryl, another metal atom, etc.), i.e.:

$$M\text{-}L + X \rightarrow M\text{-}X\text{-}L \quad (9)$$

or

$$M\begin{matrix}\nearrow X \\ \searrow L\end{matrix} \rightarrow M\text{-}X\text{-}L \tag{10}$$

Some specific examples of such reactions are:

$$[H_3CMn(CO)_5] + PPh_3 \longrightarrow [H_3C-\overset{\overset{\displaystyle O}{\|}}{C}-Mn(CO)_4(PPh_3)] \tag{11}$$

$$[HPt(PEt_3)_2Cl] + CH_2{=}CH_2 \longrightarrow [CH_3CH_2PtCl(PEt_3)_2] \tag{12}$$

$$[HMn(CO)_5] + CH_2{=}CH-CH{=}CH_2 \longrightarrow [CH_3CH{=}CHCH_2Mn(CO)_5] \tag{13}$$

Such reactions constitute essential steps in many catalytic processes such as carbonylation and addition reactions to olefins, e.g. hydrogenation, polymerization, etc..

ORIGINS OF CATALYTIC ACTIVITY OF METAL IONS AND COMPLEXES

Catalysts function by providing new paths for chemical reactions, the contributions from which are reflected in increased reaction rates. Sometimes such catalytic paths are closely related to those which operate in the absence of the catalysts; more generally, the catalytic path is a distinctive one, into which the catalyst enters as a reactant, undergoes chemical transformation, but is ultimately regenerated, so that its concentration remains undiminished.

For any given catalytic path there can usually be constructed a corresponding uncatalysed reaction path, which may or may not be an important contributing path for the actual uncatalysed reaction. In this context, the role of the catalyst may be understood as that of stabilizing the intermediate states of the corresponding uncatalysed reaction path, this stabilization necessarily being greater than that of the reactants.

An illustration of this theme is provided by the catalysis by Cu^{2+} and its complexes of the oxidation of molecular hydrogen by various oxidants, among them chromium(VI), iron(III), thallium(III) and iodate. This catalysis is effected through the mechanism, depicted by equations (14-16), in which the rate-determining step is the heterolytic splitting of H_2 by Cu^{2+}:

$$Cu^{2+} + H_2 \xrightarrow{\text{slow}} CuH^+ + H^+ \quad (\Delta H^{\ddagger} = 26 \text{ kcal/mole}) \tag{14}$$

$$CuH^+ + Tl^{3+} \xrightarrow{\text{fast}} Cu^{2+} + H^+ + Tl^+ \tag{15}$$

$$H_2 + Tl^{3+} \longrightarrow 2H^+ + Tl^+ \tag{16}$$

The corresponding uncatalysed reaction path, depicted by equations (17) and (18),

$$H_2 \longrightarrow H^- + H^+ \quad (17)$$

$$H^- + Tl^{3+} \longrightarrow H^+ + Tl^+ \quad (18)$$

$$H_2 + Tl^{3+} \longrightarrow 2H^+ + Tl^+ \quad (19)$$

is much less favorable, since the endothermicity of reaction (17) and, hence, its activation energy, can be estimated to exceed 35 kcal/mol, as compared with $\Delta H^{\ddagger}$ = 26 kcal/mol for the Cu^{2+}-catalysed reaction. This lowering of the activation energy can be attributed to stabilization by the catalyst, Cu^{2+}, of the intermediate H^-. Other instances of catalysis may be similarly interpreted.

Metal ions and complexes play important catalytic roles in a variety of homogeneous, heterogeneous and biological reactions, including oxidation-reduction processes, hydrolytic reactions, substitution reactions, reactions of saturated molecules such as hydrogen, and a great variety of addition and isomerization reactions of unsaturated molecules such as olefins. Contributing to this extraordinary catalytic versatility, the full scope of which has come to be appreciated only within the last few years, are a number of clearly recognizable factors. These include:

(1) The ability of a positively charged metal ion to increase the positive charge density on a substrate, thereby rendering it more susceptible to nucleophilic attack and thus to catalyse hydrolytic and related reactions.

(2) The ability of transition metals to stabilize a great variety of ligands through coordination. Among these are σ-bonded ligands such as H^- and alkyl groups, as well as a variety of π-bonded ligands such as olefins, acetylenes, polyenes, allyl groups, di-imine, etc..

(3) The accessibility of different oxidation states and coordination numbers, a feature not entirely restricted to, but most generally exhibited by, coordination compounds of transition metals. This property is particularly relevant to the catalytic role of such compounds in redox reactions.

(4) The existence of relatively stable but highly reactive complexes of transition metals, notably low-spin five-coordinate d^7 (17 valence electron) and four coordinate d^8 (16 valence electron) complexes which, by virtue of their open-shell electronic structures, exhibit high degrees of reactivity reflected, for example, in oxidative addition reactions such as those discussed earlier (see, for example, equations (7) and (8)).

ROLES OF METAL IONS AND COMPLEXES IN CATALYSIS

Catalysis of Nucleophilic Reactions by Electron Withdrawal from Reactants

Coordination to a positively charged metal ion is expected to increase the positive charge density on a substrate, thereby rendering it more susceptible to nucleophilic attack. The role of a metal-ion catalyst in such cases is essentially that of a Lewis acid or 'superacid' and the origin of such catalytic activity is closely related to that underlying many of the familiar catalytic effects of Brönsted acids.

In comparison with protons, metal ions, despite the higher positive charges which they can bear, are frequently not as effective, because of their more ionic bonding (particularly when coordinated to small electronegative atoms such as oxygen or nitrogen), in polarizing, or transferring positive charge to, coordinated substrates. This is reflected, for example, in the relative acidities of protonated and coordinated acids, some values of which are listed in Table I.

TABLE I

Comparison of Acidities of Free (XH_n)*, Protonated* (XH_{n+1}) *and Coordinated* (MXH_n) *Acids*

Free acid	pK_a	Protonated acid	pK_a	Coordinated acid	pK_a
H_2O	15.7	H_3O^+	-1.7	$[Li(H_2O)_n]^+$	14
				$[Be(H_2O)_n]^{2+}$	6
				$[Ni(H_2O)_n]^{2+}$	9
				$[Fe(H_2O)_n]^{3+}$	3
				$[U(H_2O)_n]^{4+}$	2
				$[Co(NH_3)_5(H_2O)]^{3+}$	6
$H_2PO_4^-$	7.2	H_3PO_4	1.1	$[Co(NH_3)_4(PO_4H_2)]^{2+}$	3.7
HPO_4^{2-}	12.7	$H_2PO_4^-$	7.2	$[Co(NH_3)_5(PO_4H)]^+$	8.5

There are, however, several circumstances in which metal ions may be more effective than protons in catalysing nucleophilic reactions. Among these are:

(a) With species (e.g. 'soft' or 'Class b' bases) such as halide ions, which exhibit low basicity toward protons, but high basicity toward certain metal ions. Promotion by metal ions of halide ion displacement (reaction (20)) from both organic and inorganic halides (a class of reactions not generally susceptible to Brönsted acid catalysis) serves to illustrate this effect. (In this and subsequent equations dashed lines indicate the intermediate steps of the proposed reaction sequences).

$$RCl \xrightarrow[H_2O]{Ag^+,\ etc.} ROH + H^+ + AgCl \qquad (20)$$

$$\left[Cl^- \cdots Ag^+ \; ; \; R^+ \; ; \; OH_2 \right]$$

$[R = CH_3, (NH_3)_5Co^{3+}, etc.]$

(b) When chelation contributes to the binding of the substrate to (and/or stabilization of the product by) the metal ion as in the following examples:

(i) Catalysis by various metal ions of the hydrolysis of amino-acid esters, according to the mechanism:

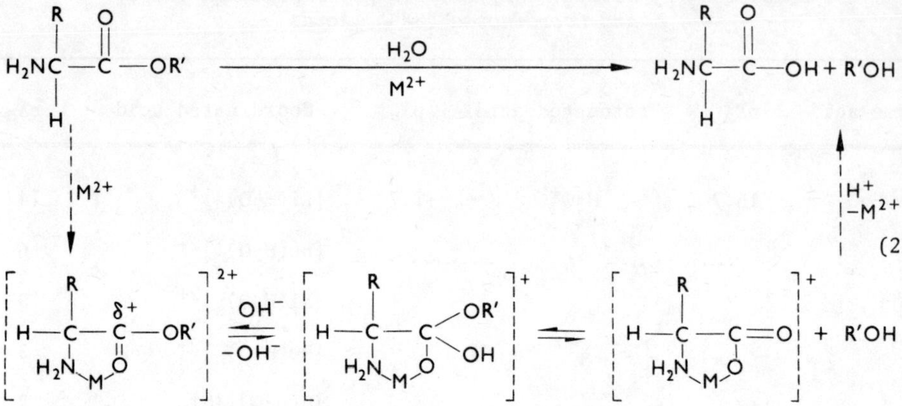

$$H_2NC(R)(H)-C(=O)-OR' \xrightarrow[M^{2+}]{H_2O} H_2NC(R)(H)-C(=O)-OH + R'OH$$

$$\downarrow M^{2+} \qquad\qquad \uparrow H^+, -M^{2+} \qquad (21)$$

$$\left[H-C(R)(H_2N)-\overset{\delta+}{C}(=O)-OR'\right]^{2+} \underset{-OH^-}{\overset{OH^-}{\rightleftharpoons}} \left[H-C(R)(H_2N)-C(OR')(OH)(O)\right]^{+} \rightleftharpoons \left[H-C(R)(H_2N)-C(=O)(O)\right]^{+} + R'OH$$

The influence of coordination in activating amino-acid esters for nucleophilic attack is also revealed by the facile N-terminal addition of amino acid or peptide esters to the coordinated glycine ester NH_2CH_2 (where R = OCH_3) in the stable complex, $[Co(en)_2(NH_2CH_2OR)]^{3+}$.

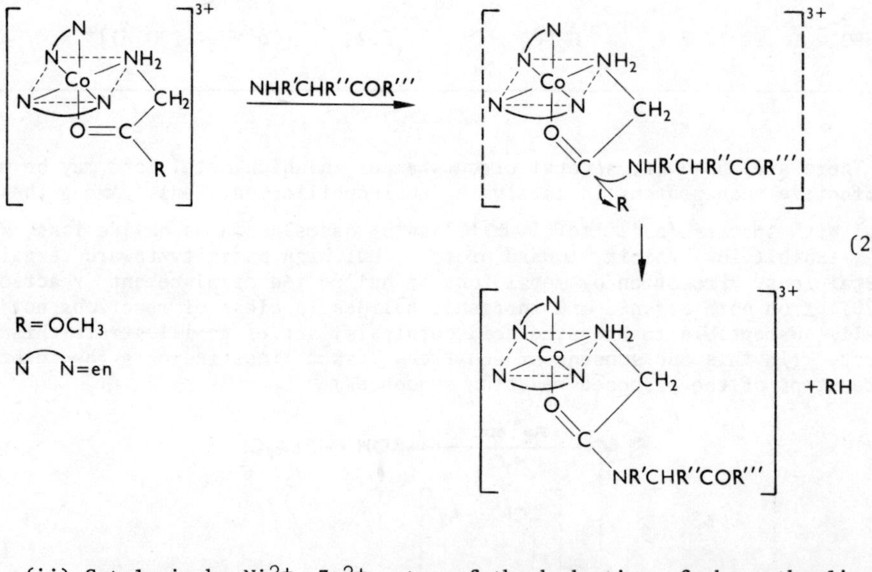

$$[Co(en)_2(NH_2CH_2C(=O)R)]^{3+} \xrightarrow{NHR'CHR''COR'''} [Co(en)_2(NH_2CH_2C(O)(R)NHR'CHR''COR''')]^{3+} \rightarrow [Co(en)_2(NH_2CH_2C(O)NR'CHR''COR''')]^{3+} + RH \qquad (22)$$

R = OCH_3

N⌒N = en

(ii) Catalysis by Ni^{2+}, Zn^{2+}, etc. of the hydration of phenanthroline nitrile. Coordination of the phenanthroline nitrile to Ni^{2+} increases the rate of attack by OH^-, according to the following proposed mechanism, by a factor of about 10^7.

$$\text{2-cyano-1,10-phenanthroline} \xrightarrow[Ni^{2+}]{H_2O} \text{1,10-phenanthroline-2-carboxamide } (C(=O)NH_2)$$

$$\downarrow Ni^{2+},\ OH^- \qquad\qquad \uparrow H^+,\ -Ni^{2+}$$

$$[\text{phen-CN}\cdots Ni^{2+} + {}^{\ominus}OH]^+ \dashrightarrow [\text{phen}-C(OH)=N^- \cdots Ni^{2+}]^+ \quad (23)$$

(c) With species such as CO, olefins, acetylenes, etc., whose coordination to metal ions depends on π-bonding through donation of $d\pi$ electrons from the metal to the ligand. Catalysis by certain metal ions such as Pd^{2+}/Cu^{2+} of the oxidation of olefins (equation (24)) by O_2 is attributable to such interaction:

$$C_2H_4 + O_2 \xrightarrow{Pd^{II}/Cu^{II}} CH_3CHO \quad (24)$$

A widely accepted mechanism for this reaction is given in equations (25) and (26).

$$Pd^{II}Cl_4 + C_2H_4 + H_2O \rightleftharpoons Pd^{II}Cl_2(CH_2{=}CH_2)(OH) + 2Cl^- + H^+$$

$$Pd^{II}Cl_2(CH_2{=}CH_2)(OH) \xrightarrow{Cl^-} [Pd^{II}Cl_3(CH_2{=}CH_2OH)] \longrightarrow$$

$$Pd^{II}Cl_3H + CH_3CHO \quad (25)$$

$$\text{Cl}_3\text{Pd}^{II}\text{H} \xrightarrow[\text{CuCl}_2]{\text{O}_2} \text{Cl}_3\text{Pd}^{II}\text{Cl} \quad (26$$

Catalysis Through Ligand Stabilization

Catalysis of Electrophilic Reactions by Promotion of Proton Loss from a Coordinated Reactant

Paradoxically (in view of their positive charges), metal ions may also catalyse *electrophilic* attack by promoting the loss of a proton from a coo inated reactant, or by stabilizing a reactive (e.g. enol) form of the latte

Catalysis by Cu^{2+} of the oxidation of molecular hydrogen (regarded as el trophilic attack on hydrogen) through a mechanism involving the heterolyti splitting of H_2 (equations (14-16)) may be considered as an example of thi effect. Other metal ions which catalyse the oxidation of H_2 by analogous mechanisms include Ag(I), Cu(I), Hg(II), Rh(III) and Ru(III). A complete description of the mechanisms of such reactions involves the recognition, (i) that coordination of the H^- ion to the metal ion involves the displacement of one of the originial ligands of the catalyst, as in the reaction,

$$[RuCl_6]^{3-} + H_2 \rightleftharpoons [RuHCl_5]^{3-} + H^+ + Cl^- \quad (27)$$

and, (ii) that, in addition to the metal ion, such catalyst systems must in volve an appropriate base to stabilize the released proton. The role of th base may be assumed by a solvent molecule or, in certain cases, by the displaced ligand, the mechanism in such cases being of the form:

$$M{-}X + H_2 \longrightarrow \begin{matrix} M \cdots X \\ \vdots \quad \vdots \\ H^- \cdots H^+ \end{matrix} \longrightarrow M{-}H^- + X{-}H^+ \quad (28)$$

The variation of rate with the ligand X, in such cases, reflects the expect direct dependence on the basicity of X, and inverse dependence on the stability of the M—X bond.

Other examples of electrophilic substitution reactions, catalysed by meta ions through mechanisms involving the promotion of proton loss and stabiliz tion of enol forms of reactant, include the bromination of β-diketones (equa tion (29)) and of ketoesters. The order of catalytic activities in the cas of reaction (29), ($Cu^{2+} > Ni^{2+} > Zn^{2+} > Mn^{2+} > Ca^{2+}$) parallels the order of the stabilities of the acetonylacetonate complexes of the metal ions.

Promotion of Ligand Synthesis by Ligand Stabilization and/or The Template Effect

The synthesis of species which are stabilized by coordination to metal ions is frequently promoted by the presence of such metal ions. Examples

$$CH_3\overset{O}{\overset{\|}{C}}CH_2\overset{O}{\overset{\|}{C}}CH_3 + Br_2 \xrightarrow{M^{2+}} CH_3\overset{O}{\overset{\|}{C}}CHBr\overset{O}{\overset{\|}{C}}CH_3 + H^+ + Br^- \quad (29)$$

of such reactions include the oxidative dehydrogenation of bipiperidine (equation (30)), the synthesis of Schiff's bases (equation (31)) and, as an extreme case, the synthesis of a molecule, namely cyclobutadiene, which is stable only as a coordinated ligand (equation (32)).

$$3\ \text{bipiperidine} + 3O_2 \xrightarrow{Fe^{2+}} [Fe(\text{bipiperidine-diimine})_3]^{2+} + 6H_2O \quad (30)$$

$$6NH_2CH_3 + 3\ CH_3COCOCH_3 \xrightarrow{Fe^{2+}} [Fe(CH_3N{=}C(CH_3)C(CH_3){=}NCH_3)_3]^{2+} + 6H_2O \quad (31)$$

$$\text{dichlorocyclobutene} \xrightarrow{Fe_2(CO)_9} C_4H_4\text{—}Fe(CO)_3 \quad (32)$$

The products of such reactions frequently remain strongly coordinated to the metal, so that the reactions are promoted but not truly catalytic. The kinetic and mechanistic information available about these reactions is insufficient to determine whether kinetic or thermodynamic factors, or both, are responsible for the observed enhancement of reactivity. This critical question also remains to be answered for a number of reactions in which it appears that catalysis or promotion of reaction results from the simultaneous coordination of two or more reactants to a metal ion in a manner which facilitates their mutual approach and favorable orientation for reaction.

Examples in which such a 'template effect' may be important include a number of reactions involving the synthesis of macrocyclic ligands, such as the metal ion-promoted synthesis of porphyrins (equation (33)), of corrins and of phthalocyanines and the cyclocondensation of *o*-aminobenzaldehyde (equation (34)).

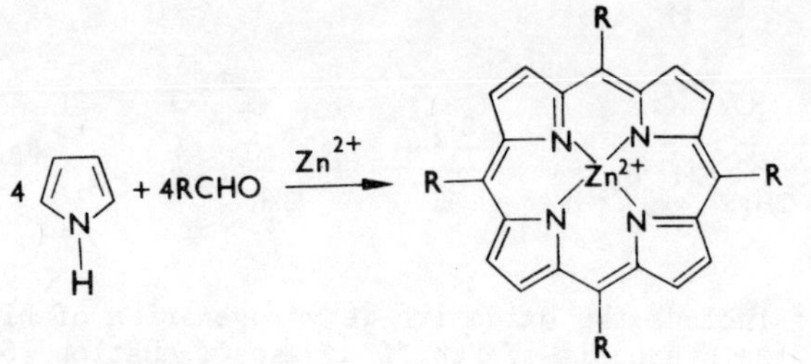

$$4\ \text{pyrrole} + 4RCHO \xrightarrow{Zn^{2+}} \text{Zn}^{2+}\text{-tetra-R-porphyrin} \quad (33)$$

$$4\ o\text{-}H_2NC_6H_4CHO \xrightarrow{M^{2+}} [M(\text{tetra-anhydro-}o\text{-aminobenzaldehyde})]^{2+} + 4\ H_2O \quad (34)$$

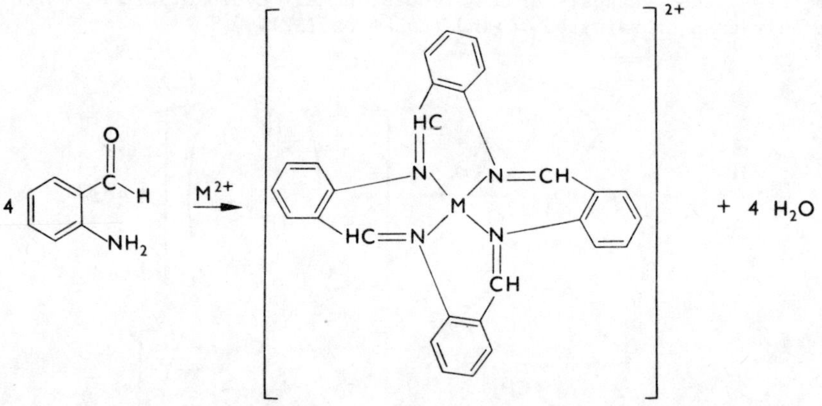

(M=Cu, Ni, or Co)

The cyclisation of allylic dibromides by nickel carbonyl, a possible mechanism of which is depicted in equation (35), represents another interesting example of a 'template' reaction:

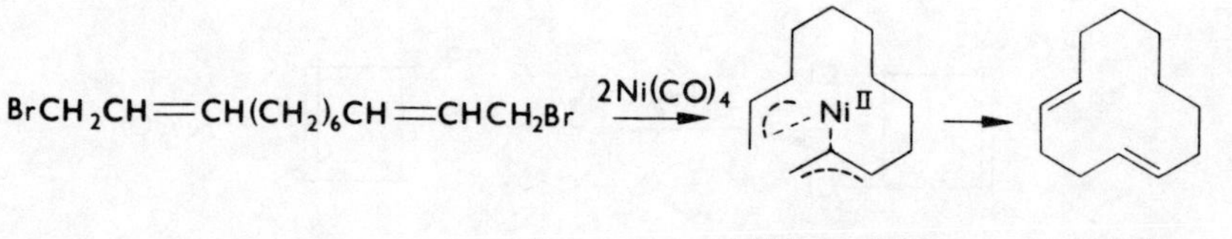

$$BrCH_2CH{=}CH(CH_2)_6CH{=}CHCH_2Br \xrightarrow{2Ni(CO)_4} \text{bis}(\pi\text{-allyl})Ni^{II} + NiBr_2 + 8CO \longrightarrow \text{cyclododecadiene} + Ni^0 \quad (35)$$

Clear-cut indications of kinetic 'template' effects are provided by a few reactions which, in contrast to those described above, are truly catalytic. Among these are the nickel(0)-catalysed trimerization of butadiene to cyclododecatriene (equation (36)) and the stereospecific dimerization of norbornadiene to 'Binor-S' (equation (37)):

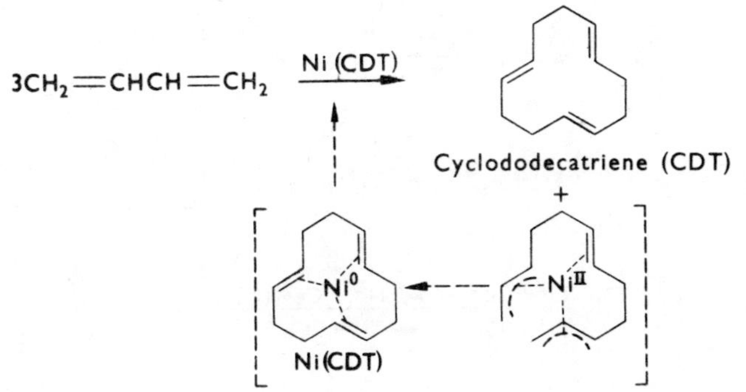

$$3CH_2{=}CHCH{=}CH_2 \xrightarrow{Ni(CDT)} \text{Cyclododecatriene (CDT)} \qquad (36)$$

Ni(CDT) Ni^0 Ni^{II}

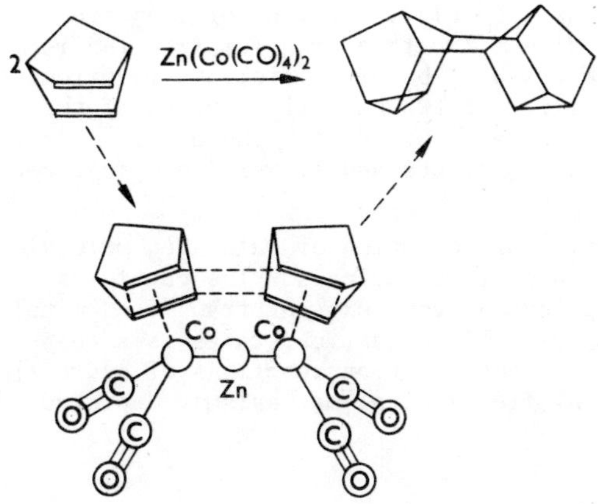

$$2 \xrightarrow{Zn(Co(CO)_4)_2} \qquad (37)$$

Co Co Zn

Catalysis of Oxidation-Reduction and of Free-Radical Reactions by Electron Transfer

The accessibility of different oxidation states for certain transition metal complexes (in some cases coupled with differences in coordination number) gives rise to catalysis of a variety of reactions, notably of the redox and free-radical types.

Among the simplest such examples are those reactions in which a metal ion or complex, having two accessible oxidation states, catalyses electron transfer by acting as an electron carrier. Two such examples are the Cu^{II}-catalysed oxidation of V^{III} by Fe^{III} (equations (38-40)) according to the rate-law, $k[Cu^{II}][V^{III}]$, and the Ag -catalysed oxidation of Tl^{I} by Ce^{IV} (equations (41-44)) according to the rate-law $k_1k_2[Tl^{I}][Ce^{IV}][Ag^{I}]/(k_{-1}[Ce^{III}] + k_2[Tl^{I}])$:

$$V^{III} + Cu^{II} \longrightarrow V^{IV} + Cu^{I} \quad \text{(rate-determining)} \qquad (38)$$

$$Cu^{I} + Fe^{III} \longrightarrow Cu^{II} + Fe^{II} \qquad (39)$$

$$V^{III} + Fe^{III} \longrightarrow V^{IV} + Fe^{II} \qquad (40)$$

$$Ag^{I} + Ce^{IV} \underset{k_{-1}}{\overset{k_1}{\rightleftharpoons}} Ag^{II} + Ce^{III} \quad (41)$$

$$Tl^{I} + Ag^{II} \xrightarrow{k_2} Tl^{II} + Ag^{I} \quad (42)$$

$$Tl^{II} + Ce^{IV} \longrightarrow Tl^{III} + Ce^{III} \text{ (fast)} \quad (43)$$

$$Tl^{I} + 2Ce^{IV} \longrightarrow Tl^{III} + 2Ce^{III} \quad (44)$$

The basis for such catalysis is that electron transfer from the reductant to the catalyst, followed by electron transfer from the catalyst to the oxidant, is faster than direct electron transfer from the reductant to the oxidant. An obviously necessary, but insufficient, requirement for such catalysis is the accessibility of two oxidation states of the catalyst, neither of which must be too stable compared with the other. For reasons which are still not well understood, copper and silver ions are especially effective as catalysts of this type.

One-electron oxidation or reduction of saturated molecules frequently results in the generation of free radicals. The catalysis of certain free-radical reactions, by ions or complexes of transition metals (such as Cu, Co and Mn) which exhibit variable oxidation states is a consequence of this. Among such reactions are autoxidation reactions of hydrocarbons (equation (45)), which proceed by free-radical mechanisms, depicted in part by equations (46-54).

Overall Reaction:

$$RH + O_2 \longrightarrow ROOH \quad (45)$$

Uncatalysed Mechanism:

Initiation:

$$RH \longrightarrow R^{\cdot} + H\cdot \quad (46)$$

$$RH + O_2 \longrightarrow R\cdot + HO_2\cdot \quad (47)$$

$$ROOH \longrightarrow RO\cdot + OH\cdot \quad (48)$$

$$RO\cdot + RH \xrightarrow[\text{etc}]{} ROH + R\cdot \quad (49)$$

Propagation:

$$R\cdot + O_2 \longrightarrow RO_2\cdot \quad (50)$$

$$RO_2\cdot + RH \longrightarrow ROOH + R\cdot \quad (51)$$

Termination:

$$2RO_2\cdot \longrightarrow \text{Inactive Products} \quad (52)$$

Mechanism of Catalysed Initiation by Co^{III}, etc.:

$$ROOH + Co^{III} \longrightarrow RO_2\cdot + Co^{II} \quad (53)$$

$$ROOH + Co^{II} \longrightarrow RO\cdot + Co^{III} \quad (54)$$

The reactions responsible for the catalytic initiation, i.e. the oxidation and reduction of ROOH, probably involve coordination of ROOH to the metal ion catalysts. Deactivation of such catalysts by chelating agents such as ethylenediaminetetraacetic acid may result from the blocking of this coordination.

Not all cases of catalysis of oxidation reactions by metal complexes involve free-radical mechanisms, e.g. catalysis of the oxidation of olefins by palladium(II) chloride (see equations (25,26)). Another example of catalytic oxidation by a non-free-radical mechanism is the catalysis of the oxidation of phosphines by Pt(0) complexes such as $Pt(PPH_3)_3$ depicted by equations (55-57):

$$[Pt(PPh_3)_3] + O_2 \longrightarrow [Pt(PPh_3)_2(O_2)] + PPh_3 \quad (55)$$

$$[Pt(PPh_3)_2(O_2)] + 3PPh_3 \longrightarrow [Pt(PPh_3)_3] + 2Ph_3PO \quad (56)$$

$$2PPh_3 + O_2 \longrightarrow 2Ph_3PO \quad (57)$$

Catalytic Applications of Oxidative Addition Reactions

Hydrogenation

Illustrative of the reactions in this class are the hydrogenation of olefins through mechanisms involving the oxidative addition of molecular hydrogen to form reactive hydrido-complexes. One such example is the $[Co(CN)_5]^{3-}$-catalysed hydrogenation of butadiene in aqueous solution according to the proposed mechanism:

$$2[Co(CN)_5]^{3-} + H_2 \longrightarrow 2[Co(CN)_5H]^{3-} \quad (58)$$

$$[Co(CN)_5H]^{3-} + CH_2{=}CH{-}CH{=}CH_2 \longrightarrow \left[(NC)_5Co\begin{matrix} CH_3 \\ / \\ CH \\ \| \\ CH \\ / \\ CH_2 \end{matrix}\right]^{3-} \quad (59)$$

$$\left[(NC)_5Co-CH_2-CH=CH-CH_3\right]^{3-} \xrightarrow{[Co(CN)_5H]^{3-}} CH_3CH_2CH{=}CH_2 + 2[Co(CN)_5]^{3-} \quad (60)$$

$$-CN \downarrow\uparrow +CN$$

$$\left[(NC)_4Co\cdots(\pi\text{-}CH_2CHCHCH_3)\right]^{2-} \xrightarrow{[Co(CN)_5H]^{3-}} CH_3CH{=}CHCH_3 + 2[Co(CN)_5]^{3-} \quad (61)$$

The CN^--dependent equilibrium between the σ- and π-allyl intermediates accounts for the observation that hydrogenation at high CN^- concentrations yields predominantly 1-butene and at low CN^- concentrations predominantly *trans*-2-butene.

An application of an oxidative addition reaction of the type of equation (8) is provided by the homogeneous catalysis of the hydrogenation of olefins by $[RhCl(PPh_3)_3]$ through the proposed mechanism shown in equation (62):

$$[Rh^{I}L_3Cl] \xrightarrow{H_2} [Rh^{III}H_2L_3Cl] \xrightarrow[-L]{CH_2=CHR} [Rh^{III}H_2L_2Cl(CH_2=CHR)] \xrightarrow{L} [Rh^{III}H(CH_2CHR)L_3Cl] \longrightarrow [Rh^{I}L_3Cl] + CH_3CH_2R \quad (62)$$

Still another mechanism of formation of reactive hydrido-complexes is through the heterolytic splitting of hydrogen as exemplified by equation (63)

$$[Ru^{III}Cl_6]^{3-} + H_2 \rightleftharpoons [Ru^{III}Cl_5H]^{3-} + H^+ + Cl^- \quad (63)$$

An example of the homogeneous hydrogenation of an olefine by such a mechanism is shown in equation (64):

$$\text{Ru}(\eta^2\text{-ABC=CBA}) \xrightarrow[\text{(slow)}]{+H_2} \text{H}^-\text{Ru}(\text{H}^+)(\text{ABC=CBA}) \longrightarrow \cdots \longrightarrow \text{Ru–C(H)AB–C(A)B} \xrightarrow{+H^+} \text{Ru} + \text{HC(A)B–C(A)BH} \tag{64}$$

$$\text{Ru} + \text{BAC=CAB} \xrightarrow{\text{fast}} \text{Ru}(\eta^2\text{-olefin})$$

This mechansim has many features in common with that proposed for the heterogeneous catalytic hydrogenation on a chromia gel catalyst (equation (65)):

$$\text{Cr}\cdots\text{O} \xrightarrow{H_2} \text{Cr(H)}\cdots\text{O(H)} \xrightarrow{\text{C=C}} \text{H–C–C(–Cr)}\ \text{H–O} \longrightarrow \text{Cr}\cdots\text{O} + \text{H–C–C–H} \tag{65}$$

Isomerization of Olefins

Complexes of many later transition metals including Co, Rh, Ir, Fe, Ni, Pd, and Pt have been found to catalyse double-bond migration in terminal olefins. Evidence for a mechanism of the following type, which is probably also applicable to some of the other catalysts, has been obtained for the rhodium chloride-catalysed reaction (equation (66)):

$$\text{Rh}^{I}L_4 \underset{L}{\overset{\text{RCH}_2\text{CH=CH}_2}{\rightleftharpoons}} \text{Rh}^{I}L_3(\text{CH}_2\text{=CHCH}_2\text{R}) \overset{H^+ + L}{\rightleftharpoons} \text{HRh}^{III}L_4(\text{CH}_2\text{=CHCH}_2\text{R}) \overset{+L,\ -L}{\rightleftharpoons} \text{Rh}^{III}L_5(\text{CH(CH}_3)\text{CH}_2\text{R})$$

$$\overset{-L,\ +L}{\rightleftharpoons} \text{HRh}^{III}L_4(\text{CH}_3\text{CH=CHR}) \overset{-(H^+ + L)}{\rightleftharpoons} \text{Rh}^{I}L_3(\text{CH}_3\text{CH=CHR}) \underset{\text{RCH=CHCH}_3}{\overset{L}{\rightleftharpoons}} \text{Rh}^{I}L_4 \tag{66}$$

(L = Cl^- or solvent)

An alternative mechanism of olefin isomerization, involving rearrangement through an intermediate π-allyl hydride (equation (67)) has also been proposed and may operate in the case of some of the other catalysts:

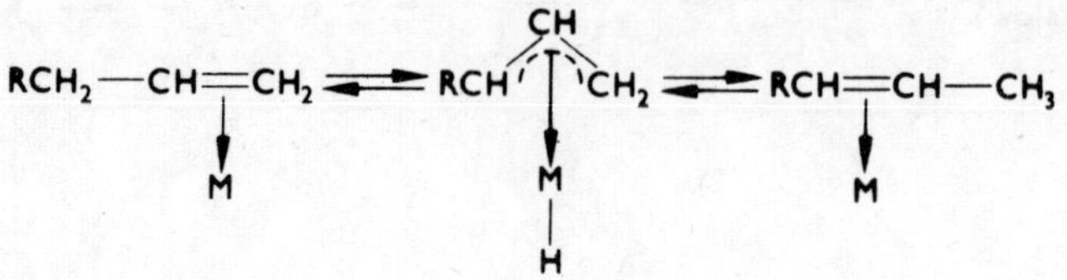

$$\underset{\downarrow \atop M}{RCH_2{-}CH{=}CH_2} \rightleftharpoons \underset{\downarrow \atop M{-}H}{RCH\overset{CH}{\cdots}CH_2} \rightleftharpoons \underset{\downarrow \atop M}{RCH{=}CH{-}CH_3} \qquad (67)$$

Dimerization of Olefins

Another olefin reaction catalysed by rhodium chloride is the dimerization of ethylene to 1-butene:

$$2CH_2{=}CH_2 \xrightarrow{Rh^{I}} CH_2{=}CHCH_2CH_3 \qquad (68)$$

A detailed study of this reaction suggests the following mechanism which is closely related to the earlier mechanism of olefin isomerization:

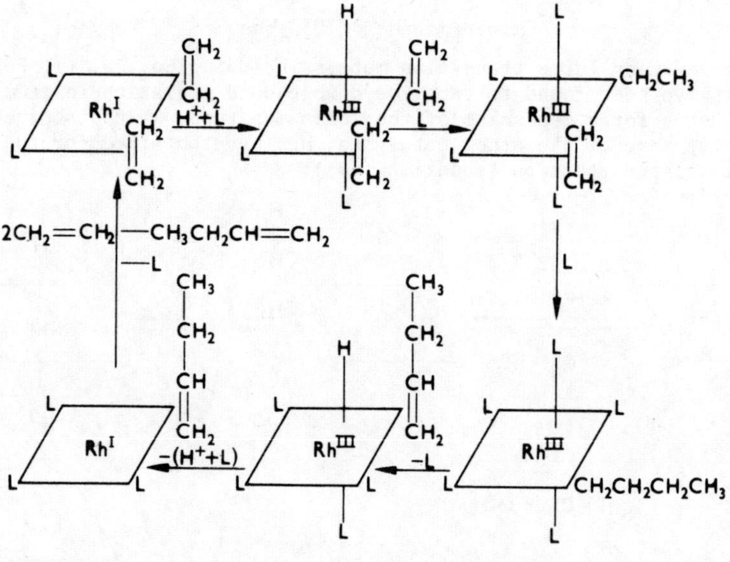

$$[Rh^{I}L_2(CH_2{=}CH_2)_2] \xrightarrow{H^{+}+L} [Rh^{III}HL_3(CH_2{=}CH_2)_2] \xrightarrow{L} [Rh^{III}L_3(CH_2CH_3)(CH_2{=}CH_2)] \xrightarrow{L} [Rh^{III}L_5(CH_2CH_2CH_2CH_3)] \xrightarrow{-L} [Rh^{III}HL_4(CH_2{=}CHCH_2CH_3)] \xrightarrow{-(H^{+}+L)} [Rh^{I}L_3(CH_2{=}CHCH_2CH_3)] \xrightarrow[-L]{2CH_2{=}CH_2,\ -CH_3CH_2CH{=}CH_2} [Rh^{I}L_2(CH_2{=}CH_2)_2] \qquad (69)$$

Hydroformylation of Olefins

The addition of H_2 and CO to olefins to form aldehydes, i.e.

$$RCH{=}CH_2 + H_2 + CO \longrightarrow RCH_2CHO \qquad (70)$$

in the presence of $[CoH(CO)_4]$ as catalyst, constitutes a reaction of great scientific interest and practical importance. The results of many investigations on this reaction, including the observation of inhibition by CO, are plausibly interpreted in terms of the following mechanism:

$$[Co^{I}H(CO)_4] \underset{+CO}{\overset{-CO}{\rightleftharpoons}} [Co^{I}H(CO)_3] \tag{71}$$

$$[Co^{I}H(CO)_3] \xrightarrow{RCH=CH_2} [Co^{I}H(CO)_3(CH_2=CHR)] \longrightarrow [Co^{I}(CH_2CH_2R)(CO)_3]$$

$$\xrightarrow{CO} [Co^{I}(CH_2CH_2R)(CO)_4] \longrightarrow [Co^{I}(-\overset{O}{\overset{\|}{C}}CH_2CH_2R)(CO)_3]$$

$$\xrightarrow{H_2} [Co^{III}H_2(-\overset{O}{\overset{\|}{C}}CH_2CH_2R)(CO)_3] \longrightarrow$$

$$[Co^{I}H(CO)_3] + RCH_2CH_2CHO \tag{72}$$

The above reaction is an example of a 'multiple insertion' process, in this case formally involving the 'insertions' of an olefin and a CO molecule into an H_2 molecule. There are numerous other examples of such homogeneously catalysed syntheses, which probably proceed through analogous mechanisms initiated by an oxidative addition step, followed by successive insertions into the resulting metal-ligand bonds of several similar or different molecules (olefins, dienes, acetylenes, CO, etc.), and terminated by reductive elimination. Nickel(O) complexes are particularly versatile catalysts for such reactions, one example being:

$$CH_2=CHCH_2Cl + CH\equiv CH + CO + CH_3OH$$

$$\xrightarrow{Ni(CO)_4} CH_2=CH-CH_2-CH=CH-COOCH_3 + HCl \tag{73}$$

Catalysis of Symmetry-Restricted Hydrocarbon Rearrangements

The rearrangement of strained hydrocarbons which are forbidden according to the Woodward-Hoffmann rules can be catalysed by a mechanism involving oxidative addition of a C-C bond as in the example shown in equation (74):

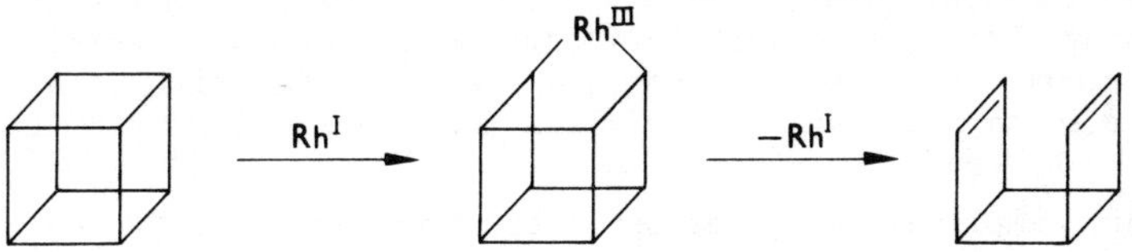

$$\text{cubane} \xrightarrow{Rh^{I}} \text{(Rh}^{III}\text{ inserted into a C–C bond)} \xrightarrow{-Rh^{I}} \text{diene product} \tag{74}$$

Another recently recognized important and highly novel class of 'symmetry-forbidden' reactions which are catalysed by transition-metal complexes, is olefin-dismutation, i.e.

$$\begin{matrix} R^1CH{=}CHR^2 & & R^1CH{=}CHR^3 & & R^1CH{=}CHR^4 \\ + & \rightleftharpoons & + & \rightleftharpoons & + \\ R^3CH{=}CHR^4 & & R^2CH{=}CHR^4 & & R^2CH{=}CHR^3 \end{matrix} \quad (75)$$

Typical heterogeneous catalysts for this reaction include Mo or W carbonyls or oxides on alumina supports, while homogeneous catalysts are derived from such combinations as WCl_6 with $C_2H_5AlCl_2$ or $Mo(NO)_2(PPh_3)_2Cl_2$ with C_2H_5AlCl The mechanisms of these reactions are not as yet understood.

Oxidative Addition Reactions of the Early Transition Metals

Certain complexes of the early transition metals, notably those of d^2 configuration, also undergo oxidative addition reactions analogous to those of the later metals. The reactivity of such complexes is often high and, thus can result in activation of aromatic C-H bonds as reflected in the catalysi by $[Nb(C_5H_5)_2D_3]$ of the exchange of C_6H_6 with D_2 according to the following scheme:

$$[Nb^{V}(C_5H_5)_2D_3] \rightleftharpoons [Nb^{III}(C_5H_5)_2D] + D_2 \quad (76)$$

$$[Nb^{III}(C_5H_5)_2D] + C_6H_6 \rightleftharpoons [Nb^{V}(C_6H_5)(C_5H_5)_2DH] \rightleftharpoons [Nb^{III}(C_5H_5)_2H] + C_6H_5D \quad (77)$$

SELECTIVITY AND STEREOSPECIFICITY

Because the chemical and structural features of homogeneous catalyst systems are often better and more reproducibly characterized than those of heterogeneous ones, and more readily susceptible to systematic control and variation (e.g. by ligand variation) it is sometimes possible to achieve much higher degrees of selectivity and stereoselectivity. This is reflected in some of the examples already cited such as the $[Co(CN)_5]^{3-}$-catalysed hydrogenation of butadiene which does not proceed beyond the butene stage. Another dramatic example is the asymmetric synthesis of optically active compounds by homogeneous hydrogenation using catalysts, notably Rh^{I} complexes, containing chiral phosphine ligands, i.e. $^{*}PR_1R_2R_3$. This approach has yielded optically active amino acids, in optical yields of up to 90%, by the catalytic hydrogenation of aminoacrylic acids. Such stereospecificity, approaching that characteristic of enzymes and rarely, if ever, achieved wit heterogeneous catalysts, represents a significant advance in synthetic metho ology for which practical applications (e.g. the synthesis of L-dopa) have already been reported.

Another example of such stereoselectivity is provided by the cyclooligomerization and cyclotrimerization of butadiene catalysed by nickel(0) complexes which yield, in reactions characterized by high degrees of selectivity and stereospecificity, a variety of products (see below) the preference for which can be controlled by the choice of ligands (e.g. different phosphines):

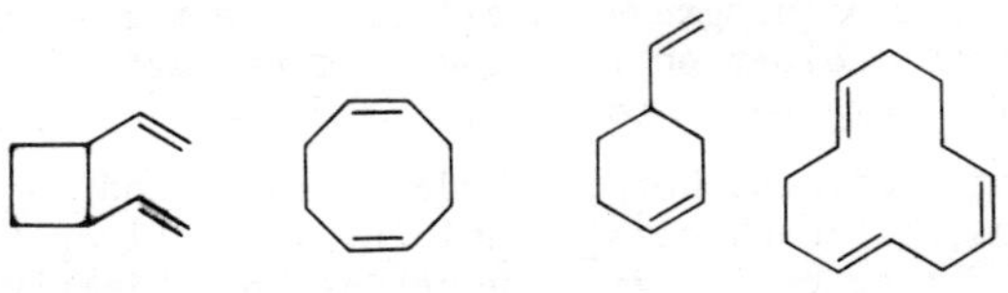

Another way in which metal ions can influence the selectivity of a reaction is by the blocking of a reactive site through coordination. This is the counterpart of the well-known and widespread applications of 'protective groups' in organic chemistry. While not catalytic in nature, this effect sometimes appears so because it results in the modification of the course of a reaction so that the formation of certain products is favored over that of others. Two such examples are the modification of the course of alkylation of cyanide by coordination to Ag(I) (equations (78,79)) and the modification of the course of bromination of an unsaturated arsine by coordination to Pt(IV) (equations (80,81)):

$$NaCN + RX \longrightarrow RCN + NaX \quad (78)$$

$$AgCN + RX \longrightarrow RNC + AgX \quad (79)$$

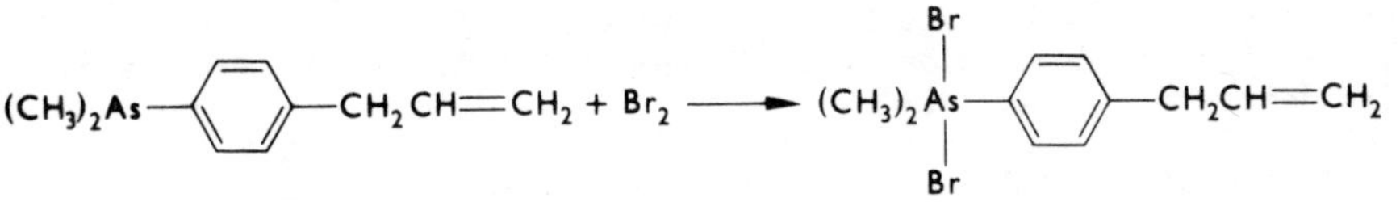

$$(CH_3)_2As-C_6H_4-CH_2CH{=}CH_2 + Br_2 \longrightarrow (CH_3)_2AsBr_2-C_6H_4-CH_2CH{=}CH_2 \quad (80)$$

$$(CH_3)_2As(\rightarrow Pt^{IV})-C_6H_4-CH_2CH{=}CH_2 + Br_2 \longrightarrow (CH_3)_2As(\rightarrow Pt^{IV})-C_6H_4-CH_2CHBr-CH_2Br \quad (81)$$

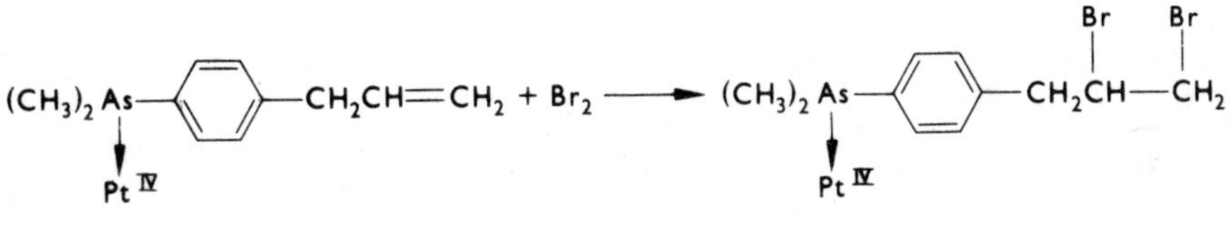

CURRENT THEMES AND UNSOLVED PROBLEMS

The modern phase of homogeneous catalysis is characterized particularly by its emphasis on catalytic systems involving metal ions and coordination compounds. Currently important lines of research are concerned particularly with the search for and discovery of new catalytic reactions, with the more detailed elucidation of the mechanisms of the many reactions whose mechanisms are as yet incompletely understood (e.g. olefin dismutation) and with the achievement of a greater mastery of the factors (such as systematic variation of electronic and steric ligand properties) that control catalytic activity, selectivity and stereospecificity. Such studies are directed not only at catalytic reactions as such but also toward the discovery and characterization of new coordination and organometallic compounds which are of interest as potential catalysts or catalytic intermediates, or whose study might contribute to a better understanding of related catalysts or catalyti intermediates. Recent progress in all of these areas has been impressive, and there is every indication that the present pace of research and discove will continue for some years to come.

Many important problems, both of understanding and application, remain to be solved. Thus, although considerable progress along these lines has already been made, many challenging questions and objectives, both of a scien tific and practical nature, continue to be associated with the catalytic ac ivation of molecular nitrogen. In contrast to the situation for certain other catalytic reactions such as hydrogenation or carbonylation, the present state of the field of nitrogen fixation encompasses relatively little understanding of the mechanisms of the several known nitrogen-fixing system and a rather limited basis for predicting new developments. Another objective, of great scientific and practical importance, which still awaits solution is the homogeneous catalytic activation of saturated hydrocarbons. Vi tually no progress toward this objective has yet been made and, while the approaches (e.g. oxidative addition) that have been successful for the catalytic activation of other saturated molecules such as molecular hydrogen and alkyl halides, should in principle be capable of extension also to saturated hydrocarbons, such extensions have yet to be achieved to a meaningful degree

SELECTED REFERENCES

AHRLAND, S., CHATT, J. and DAVIES, N.R. (1958). The relative affinities of ligand atoms for acceptor molecules and ions, *Quart. Rev. Chem. Soc.*, 12, 265.

BENDER, M.L. (1971). *Mechanisms of Homogeneous Catalysis from Protons to Proteins*, (John Wiley and Sons, New York).

BIRD, C.W. (1967). *Transition Metal Intermediates in Organic Synthesis*, (Academic Press, New York).

CALDERON, N. (1972). Olefin metathesis, *Accounts Chem. Res.*, 5, 127.

CHATT, J. and LEIGH, G.J. (1972). Nitrogen fixation, *Chem. Soc. Revs.*, 1, 121.

CHIUSOLI, G.P. (1970). Stereoselectivity in organic synthesis involving nickel-coordinated intermediates. In *Aspects of Homogeneous Catalysis, Vol. I*, (ed. Ugo, R.), (Manfredi, Milan), p. 77.

COFFEY, R.S. (1970). Recent advances in homogeneous hydrogenation of carbon-carbon multiple bonds. In *Aspects of Homogeneous Catalysis*, (ed. Ugo, R.), (Manfredi, Milan), p. 13.

HALPERN, J. (1968). Homogeneous catalysis by coordination compounds, *Adv. Chem. Ser.*, 70, 1.

HALPERN, J. (1969). Coordination compounds in homogeneous catalysis, *Pure Appl. Chem.*, 20, 59.

HALPERN, J. (1970). Oxidative addition reactions of transition metal complexes, *Accts. Chem. Res.*, 3, 386.

HALPERN, J. (1972). Catalysis of the rearrangements of strained hydrocarbons by transition metal compounds *Proc. 14th Int. Conf. Coord. Chem., Toronto*, p. 698.

HALPERN, J. (1968). Homogeneous catalysis of hydrogenation, oxidation and related reactions, *Disc. Faraday Soc.*, 46, 7.

HEIMBACH, P., JOLLY, P.W. and WILKE, G. (1970). π-Allylnickel intermediates in organic synthesis, *Adv. Organomet. Chem.*, 8, 29.

HOOGZAND, C. and HUBEL, W. (1968). Cyclic polymerization of acetylenes by metal carbonyl compounds. In *Organic Syntheses via Metal Carbonyls, Vol. I*, (eds. Pino, P. and Wender, I.), (Wiley, New York), p. 343.

JONES, M.M. (1968). *Ligand Reactivity and Catalysis*, (Academic Press, New York).

KEIM, W. (1971). π-Allyl system in catalysis. In *Transition Metals in Homogeneous Catalysis*, (ed. Schrauzer, G.N.), (Marcel Dekker, New York), p. 59.

KWIATEK, J. (1971). Hydrogenation and dehydrogenation. In *Transition Metals in Homogeneous Catalysis*, (ed. Schrauzer, G.N.), (Marcel Dekker, New York), p. 13.

LAPPERT, M.F. and PROKAI, B. (1967). Insertion reactions of compounds of metals and metalloids involving unsaturated substrates, *Adv. Organomet. Chem.*, 5, 225.

LEFEBVRE, G. and CHAUVIN, Y. (1970). Dimerization and co-dimerization of olefin compounds by coordination catalysis. In *Aspects of Homogeneous Catalysis, Vol. I*, (ed. Ugo, R.), (Manfredi, Milan), pp. 107-203.

MARTELL, A.E. (1968). Catalytic effects of metal chelate complexes, *Pure Appl. Chem.*, 17, 129.

ORCHIN, M. and RUPILIUS, W. (1972). On the mechanism of the oxo reaction, *Catalysis Revs.*, 6, 85.

PAULIK, F.E. (1972). Recent developments in hydroformylation catalysis, *Catalysis Revs.*, 6, 49.

PEARSON, R.G. (1966). Acids and bases, *Science*, 151, 172.

STERN, E.W. (1968). Reactions of unsaturated ligands in palladium(II) complexes, *Catalysis Revs.*, 1, 73.

THOMPSON, D.T. and WHYMAN, R. (1971). Carbonylation. In *Transition Metals in Homogeneous Catalysis*, (ed. Schrauzer, G.N.), (Marcel Dekker, New York), p. 147.

TOLMAN, C.A. (1972). The 16 and 18 electron rule in organometallic chemistry and homogeneous catalysis, *Chem. Soc. Revs.*, 1, 337.

TSUJI, J. and OHNO, K. (1969). Decarbonylation reactions using transition metal compounds, *Synthesis*, p. 157.

PART IV
WORKING GROUP REPORTS

KINETICS AND MECHANISMS

CHAIRMAN: W. K. Hall

Gulf Research & Development Company, Pittsburgh, Pennsylvania 15230, *USA*

RECORDER: P. Henry

Department of Chemistry, Guelph University, Guelph, Ontario, Canada

MEMBERS: S. J. Benkovic J. Bjerrum R. C. Bray
G. L. Eichorn D.D. Eley J. J. Fripiat
H. Knözinger P. M. Maitlis A. Rigo

INTRODUCTION

Before attempting to suggest new approaches to the study of the three types of catalysis it seems best to summarize the overall state of affairs. First, it should be emphasized that special problems exist which are not present in the studies of simple systems. Thus, the state of the metal or enzymatic species in solution must be specified and this often requires elaborate studies. Even when the studies have been carried out in the best possible fashion, only the slow step of the catalytic reaction is defined. Fortunately, other information such as stereochemical results, tracer studies, product distributions and studies on related systems will often be sufficient to suggest the catalytic cycle.

Homogeneous catalysis is the simplest system and this, in principle, should give the most meaningful results. Next comes enzyme catalysis, for which kinetics have been extensively employed but interpretation is more difficult than for homogeneous catalysis.

Of the three fields, kinetic studies in heterogeneous catalysis are least useful in providing information on mechanism. It is relatively easy to determine gross kinetics, but only in relatively simple cases have mechanistic interpretations on this basis been of use, e.g. in carbonium ion catalysis. Thus, the Langmuir-Hinshelwood model considers energetically identical sites, but it is known that a distribution of site energies is generally present, e.g. on different crystal faces or at lattice defects. Furthermore, the chemical kinetics do not directly measure the surface kinetics which may be the most important mechanistic feature.

Hence, because of the complexity of the systems, kinetics provide only a zeroth-order approximation to the mechanism or serve to eliminate certain mechanistic possibilities. For this reason other methods such as tracers, infrared and other spectroscopic techniques for studying surfaces have been largely used to infer reaction intermediates. Although much progress has been made, considerable work remains to be done.

It is important to define in advance of a mechanistic study what questions must be satisfactorily answered before the mechanism is said to be known. The following should be answered at a reasonable level of refinement. What is the reaction scheme, as derived from chemical evidence or tracer experiments? What are the reacting species? What are their relative concentrations? What are the modes by which they interact? What are the relative values of the rate constants of the unit steps, and which is therefore rate limiting? Given the last, what is the true activation energy of the reaction and what does the reaction coordinate look like? As one proceeds to answer these questions, one at a time, the mechanism becomes known with increasing sophistication. New theoretical approaches are needed. In particular, means of predicting kinetics from reaction schemes would be helpful. Recognition should be given to possibilities afforded by statistical design of simple and multifactor experiments and to computer modelling. In this way, it is easy to test many kinetic models. Furthermore, the kinetic and thermodynamic information required to discriminate between alternatives becomes evident.

HETEROGENEOUS CATALYSIS

General

Since mechanisms are not readily deduced from kinetics, other techniques must be applied in the hope of getting enough of the pieces together to see what the puzzles must look like. Some may be obtained without regard to surface chemistry; others must be deduced from surface studies, usually by spectroscopic means. In the former case, tracers are usually employed, as for example by adding a radioactive compound which may form an intermediate to the reactant. The distribution of the label is then determined in the products. Alternatively, specifically labelled deuterated compounds are sometimes used and the products examined to see whether the label is lost or retained. Another approach has been to study the exchange of reactant with D_2 in an attempt to deduce how it must have been bound to the catalyst. Such experiments usually provide enough insight to generate a reaction scheme which then may be compared with the kinetics to see if a match can be obtained. Frequently, it cannot and recourse has to be made to supplemental methods. These include the use of 'model compounds' or the development of linear free energy relations to answer specific questions concerning the mechanism, as well as spectroscopic measurements designed to identify or suggest possible reaction intermediates. The first two of these have so far been used effectively by only a few workers, but with promising results. More work of this kind should be encouraged.

Kinetic and equilibrium isotope effects can sometimes be used to advantage When these are large enough to be meaningful, the bond which is ruptured in the rate-determining step can be located. Many times additional insight can be obtained by looking for this effect in the reverse reaction, when this is possible.

Kinetic formalities exist between equations derived on the basis of quite different models. For example, the Michaelis-Menten and Langmuir equations are mathematically identical. Many cases of simple kinetics can be shown to result from a limiting case of one of these. Research directed towards identifying the type of process which is actually occurring should be a part of any kinetic study. Similar research is needed on systems for which kinetic data have already been published.

Techniques

To date, among the various spectroscopic tools, IR has proved the most useful, although the results are only qualitative. Such measurements could be made much more valuable if they could be made quantitative, through determination of absorption coefficients.

These techniques may be arranged according to their time resolution. Thus, IR and Laser-Raman spectroscopy, as well as UV and visible spectroscopy, may provide information concerning the nature and structure of adsorbed species. In some cases, absorption bands may be identified with reactive intermediates if the lifetime is longer than approximately 10^{-13} s. Laser-Raman spectroscopy for this application has potential for obtaining data in the low-frequency range where transmission is very low because of absorption by the catalyst itself. Nevertheless, studies in the low-frequency region by IR, where possible, should be encouraged. Recent developments of this technique should make its application to studies of the surface-adsorbate bonds possible; thus one may be able to measure the frequencies of vibrations relative to the surface. The applicability of these spectroscopic methods to metal catalysts and to simple crystal surfaces, in particular, should be developed further. Additional information can be obtained by the parallel use of flash-filament and temperature-programmed desorption experiments.

High-resolution NMR has been applied to identify adsorbed species. This approach, however, seems to be quite restricted. On the other hand, the determination of relaxation times by means of broad-line NMR and pulsed NMR techniques with a time scale in the range of 10^{-4} to 10^{-10} s has proved to have a very high potential. Pulsed NMR techniques enable one to determine local surface diffusion coefficients and their distribution. The latter is brought about by surface heterogeneity, which in turn can be related to the dependence of heats of adsorption on surface coverage. A major difficulty in the interpretation of heterogeneous kinetics is the fact that heats of adsorption fall with coverage and it is not known exactly how this should be accommodated in the kinetic equations.

In cases where free radicals are formed, ESR *may* become a useful tool. In such cases, information about motional freedom of the radical species may be obtained from the hyperfine structure of the ESR signals.

Though seldom applied, microwave and dielectric-relaxation methods may give information concerning slow molecular motions such as hindered rotations. Microwave spectroscopy furthermore is a very useful tool for analysis of deuterated products from reactions with labelled compounds.

Very little is presently known concerning the electronic energies of adsorbed molecules. ESCA potentially can improve our knowledge in this field. One of the unsolved problems, however, is the effect of charging on the electronic energy levels which occur with insulators.

Changes in surface coordination on adsorption occur, particularly with oxide systems. The application of ESCA, electronic and X-ray fluorescence spectroscopy and the determination of radial electron distributions should be useful in understanding this problem.

All the techniques mentioned above and even others may be applied at elevated temperatures to get information concerning the behavior of the adsorbed phase under reaction conditions. Recently Fourier-transform spectroscopy has become available. Infrared emission spectroscopy may be used fruitfully in the future to study reacting systems, even at fairly low temperatures. Mössbauer spectroscopy will find application.

Relaxation kinetics have been fruitfully applied in the other two fields. It is suggested that they be further developed for heterogeneous catalysis. These methods are usually coupled with a spectroscopic technique such as IR or visible-UV spectroscopy. New developments in Fourier-transform NMR afford attractive possibilities. It is easy to see how work in this area could lead to a better understanding of the activation process. However, methods are also needed which will provide information concerning rates of elementary steps during the steady-state reaction. This may prove more difficult. Although it is easy to perturb the system by sudden changes in pressure or possibly temperature, it may not prove as easy to get a sufficient number of 'handles' to solve the problem.

More information is needed concerning the characteristics of 'clean' surfaces and of the surfaces under working conditions. Any of the tools mentioned above may be applied for this purpose. Another area of possible future importance concerns the use of isolation matrix techniques to capture small clusters of metal atoms for chemisorption studies. The adsorbate coul then by 'activated' photolytically or changes which occur on raising the tem perature within the allowed limits could be assessed.

Activation

The fate of a molecule in the adsorbed phase will bring it from time to time by an intricate game of spatial and energetic fluctuations into contact with an active site, at the right moment and with the right configuration fo activation. The reactant activation is achieved by an exchange of matter an energy with the active site. Establishment of such a mechanism requires the knowledge not only of surface diffusion coefficients, and therefore of the residence time of the reactant near the site, but also of the energetic and configurational fluctuation at the active site. For instance, in the hydrogen zeolite, the proton is not always at the right position (correctly oriented in the zeolitic cavities) and its life-time in the right orientation is rather short. Moreover, these constitutional protons may be delocalized onto the four oxygens of the first co-ordination sphere of a tetrahedral cation in such a way that any of them can become an active site, although only those which have access to the reactant will act as active sites. The fluctuation of the proton between the adsorbate and the surface site may contribute to the strong polarization of the former, e.g. as in carbonium-ion formation. Similar considerations would apply to electron transfer. Fluctuations in the adsorbed phase may be at the origin of ordering the adsorbed molecules. Such order may be an important factor for the fast transmission of 'chemical information' over considerable distances. For instance, imagine an ordered structure of H_2O molecules on a surface. A fluctuation may lead to transfer of a surface proton onto an adjacent H_2O molecule. Through a co-operative effect, a proton may become available at the opposite side of the cluster at a distance, e.g. of 10 Å, within a very short time, e.g. 10^{-13} s. This kind of information may be derived from relaxation techniques such as pulsed NMR, ESR, etc..

In several instances, the polarization of the adsorbed species by a strong surface electrical field has been invoked as a possible source of activation. Evaluation of surface electrical fields of insulator surfaces may be derived from the solution of Schrödinger's equation in which a perturbation Hamiltonian accounting for these fields has been introduced. Effective fields may be estimated by their effects on oscillators (IR and Raman spectroscopy) and also by measuring NMR relaxation times of quadrupole-bearing molecules. Heat of adsorption of simple diatomic molecules and *ortho-para*-hydrogen separation factors have been used for this purpose.

Intermediates

Reaction intermediates necessarily are metastable species. Therefore, there is always a hazard that those which are identified spectroscopically are extraneous to the mechanism. To prove that they are truly intermediates, it is necessary to show that they respond to changes in external variables in the same way as the reaction rate. This has been done only in two cases, i.e. for the hydrogenation of C_2H_4 over ZnO and for CH_3OH decomposition over alumina. Much more work in this vein is needed.

Poisons and Promotors

Selective poisoning of active sites on oxides has been used in a few cases to count the number of sites which are effective for a reaction. Similarly, attempts have been made to divide the adsorbed phase into components, one of which is presumed to be on the active site, by using flash desorption techniques. Reasonable agreement has been found. With metals the entire surface is usually presumed to be active (for facile reactions), although this may not be so. Information of this kind is especially valuable because it allows reduction of the rate data into turnover numbers which in turn allows cross comparison for comparison of grossly different reaction systems and sets targets for future research. Efforts should be made to unify data from various laboratories in this manner. If this were done, it would point out inconsistencies in experimental work, e.g. where impurities have acted as poisons. Problems of this kind occur in all three fields. Various promotors are used in practical catalysts to improve activity, lifetime, or to improve selectivity. Generally, the way in which these function is incompletely understood. Use of modern new tools to investigate such problems should prove rewarding.

Selectivity

Modern chemical industry makes its profits largely by developing and using catalysts which accomplish specific ends. Sometimes this requires building more than one function into the catalyst. In the hydroisomerization or hydrocyclization of paraffins, both a hydrogenation-dehydrogenation and an acid function are required. It appears likely that future industrial catalysis will require still more selective catalysts. Moreover, it is thought that the great promise of the present decade lies in the development of highly selective catalysts by heterogenizing selective homogeneous catalysts so that the advantages of both methods may be combined. Enzymes attached at one terminus to solids may also play a role.

The factors which govern selectivity are presently poorly understood and much work in this area is indicated. Kinetic measurements define selectivity, but in most cases it is not known whether these results stem from changes in the activation energy or in the pre-exponential factors. In carbonium-ion catalysis, the former appears to be the case and a start has been made in relating selectivity to catalyst composition. This field needs considerably more attention. In the area of metal catalysis, relations are being sought between selectivity and particle size as an approach to the general problem of surface configuration *vis-à-vis* selectivity. Situations have been found where either there is an effect or there is not and these have been classified as demanding or facile reactions, respectively. The real causes of this behavior are at best poorly understood. Alloying appears to afford some attractive possibilities for research. For example ESCA, Auger, ISS (Ion-Scattering Spectroscopy) and SIMS (Secondary-Ion Mass Spectroscopy) may be used to measure alloy surface compositions. If the reaction sites are localized on a transition metal, an estimate of cluster size may be made and correlated with selectivity.

KINETICS AND MECHANISM OF HOMOGENEOUSLY CATALYSED REACTIONS

The basic catalytic process can be described by four stages, any one of which may involve either one or more steps:

(a) The generation of the 'true catalyst' from the starting complex (e.g. by loss of a ligand to create a vacant site).

(b) Activation by the metal of the substrate and, in many cases, of the reactant.

(c) Reaction in the coordination sphere to give products.

(d) Regeneration of the 'true catalyst'.

All these points need attention; (a,b,d) are largely problems in basic co-ordination chemistry while (c) is of particular relevance to catalysts.

Basic Coordination Chemistry

This is fairly well understood but a number of general problems remain; for example the nature of the metal ions in various solvents (solvation numbers), their states of aggregation, and their co-ordination numbers. More information is needed about substitution processes, particularly for labile metals and for unusual coordination numbers. The question of whether metals react by dissociative mechanisms in which vacant sites are created or whether they react by associative mechanisms in which the metal can increase its co-ordination number is of great relevance. In this connection further studies of the second coordination sphere will also be very useful. This will help to answer the problem of whether a given reaction occurs by attack of co-ordinated or uncoordinates species.

There is also very little basic information on equilibria involving catalytically interesting compounds and on the bond energies of the types of bonds involved in catalytic processes. Studies on model systems will be very helpful here.

The interpretation of kinetic data for catalytic reactions is dependent upon a knowledge of related thermodynamic data, not only about the overall reaction, but also about the interaction of reactants and products with the catalyst and about the reaction intermediates. The need for such data is very great, especially for organometallic compounds which are of interest as catalysts or catalytic intermediates. Although most of the reactions of current interest involve the formation and dissociation of metal-carbon or metal-hydrogen bonds, very few such bond energies are available. Because of lack of reversibility in many cases, equilibrium and electrochemical measurements have only limited applicability; therefore thermochemical measurements are urgently needed.

The considerable current interest in the following problems should be maintained:

(a) The effects of metal-atom clusters in catalytic reactions.

(b) The effects of changing oxidation state and coordination number.

(c) The effects of ligand and solvent changes on selectivity.

(d) The heterogenizing of homogeneous catalysts.

So far there has been a tendency to concentrate attention on the Group VIII transition metals and one or two others (Ti, Cr, Mo, W, Ca). Other metals should be investigated including main group, lanthanide, and possibly even the actinide metals.

Mechanistic Studies

Relatively few detailed mechanistic studies have been carried out on homogeneous catalysts and more are essential. The measurement of kinetics will give some help, particularly since the evaluation of E_a and $\Delta S^{\ddagger}$ should shed some light on the basic processes occurring. They should also help our understanding of the kinetic and thermodynamic factors involved in product selectivity.

Kinetics will, at best, give only information on the slowest step. Catalytic reactions are multistep and in many cases the step for which information is required is very fast. Various ways of overcoming these difficulties are available: isotopic labelling and the investigation of the stereochemistry of the reaction; alteration of ligands or solvent to alter the relative rates of reactions, and, particularly, the study of model systems which may either involve the same metal atom or a kinetically more inert congener.

New instrumentation should also help to detect and identify intermediates. One area which has been little explored is the use of stopped-flow techniques and the development of a device allowing single-pulse Fourier-transform NMR monitoring in a stopped-flow or similar reaction. Such experiments would give direct and detailed information concerning the natures and even the conformations of the species present. Other tools with higher sensitivity should also be utilized more extensively, e.g. ESR on paramagnetic metal complexes or spin-labelled ligands. The possibility of freezing out reactions and then examining the constituents should also be examined.

Theoretical Studies

Although, in general, the systems being dealt with are too complex for very refined calculations it would be of great interest to carry out approximate studies (CNDO, for example) to shed light on the basic processes involved and on the transition states. Two examples are the nature of the activation process and the *cis*-ligand migration reaction.

Other Studies

The building of model systems to approximate both enzymatic and surface catalysis is of major importance. Model enzymatic systems need the flexibility characteristic of known enzymes as well as the presence of functional groups on the ligands to help bind the substrate. It should be possible to mimic enzyme behavior as well as to tailor make 'enzymes' for specific processes.

The most reasonable analogue for a heterogeneous metal catalyst is a cluster complex. Most of those known today are disappointing catalytically and effort should be expended on developing more labile systems. One particular problem this could help to resolve is the relative importance of single versus multiple sites.

A logical extension of both these studies is the development of multifunctional catalysts comprised of two or more metals — or of one metal with auxiliary binding sites. This would allow the development of new processes, particularly for stereospecific syntheses.

ENZYMATIC CATALYSIS

Over the last decade very large advances have been made in our understanding of the kinetics and mechanisms of enzymatic catalysis. Much of the present effort is hampered by the continuing difficulty of preparing large quantities of pure enzymes. A great deal of our present understanding of the mechanism of enzyme action is based on detailed knowledge of enzyme structure derived from X-ray crystallography; hence, the emerging picture i principally a static one. Ideally one would like a complete kinetic descri tion of the individual steps, the conformational changes which presumably connect them and the nature of the intermediates involved. It is clearly desirable to extend the applicability of present methods of studying enzyme in solution by improving the sensitivity of existing physical measurements such as NMR augmented by the synthesis of specifically isotopically enriche substrates.

Perhaps these techniques might be combined with T-jump or stopped-flow ki etic methods to define more precisely the reaction sequence and to detect catalytically labile intermediates. Measurement of kinetic isotope effects should also prove a useful tool in this regard. Physical procedures with lower time resolution may be applicable in those cases where intermediates may be trapped, e.g. by sudden freezing. The importance of understanding th nature of conformational changes representing an integral part of the catalytic process cannot be overemphasized. Hence improvement of relevant methodology, e.g. fluorescence, as well as theory, will clearly be important.

Although considerable attention has been given to the factors underlying enzymatic catalysis, many questions remain unanswered. The problem of rate enhancements, particularly in regard to the metalloenzymes, is very poorly understood; at least in part this is due to the absence of studies on appropriate model systems. More specific problems include the role of solvent interaction, e.g. the structural water within the active site of carbonic anhydrase; and electron transfer to and between metals buried within individual proteins and between the metals of distinct proteins in coupled systems, e.g. as with membranes. The kinetics of the organized system of enzymes also pose problems for future research.

AREAS OF OVERLAP

The first requirement is that absolute rates, or turnover numbers in molecules per site per second should be available for comparison between comparable heterogeneous, enzymatic, and homogeneous catalysis. The main difficulty here concerns estimating the number of active sites in *heterogeneous* catalysts, but similar problems may also exist in specific homogeneous catalysts and enzymes, especially membrane-bound enzymes.

There is a general point that kinetics, however thoroughly determined over the widest range of temperature, concentration of reactants, etc., cannot alone lead to unequivocal mechanisms in the absence of knowledge of catalyst structure and reaction intermediates. Concerning kinetics, measurement of pre-steady-state rates which is now conventional in enzyme and homogeneous work could be usefully extended to suitable heterogeneous systems. The word *activation* has been used in two senses — activation of the substrate (reactants) in heterogeneous and homogeneous catalysis, and activation of the enzyme by additional entities — e.g. metals of coenzymes. In either case the process leads to labile intermediate species, and the identification of these by spectroscopic methods is common to all three branches. Common also is the difficulty of deciding whether a majority species observed in fact lies in the reaction path, since by definition catalytic intermediates are usually present in lower concentrations. A possible development of high-resolution scanning microscopy is the identification of active intermediates or inhibito complexes for enzymatic and surface reactions. Kinetic considerations may be dominated by diffusion of reactants or products in the case of heterogeneous catalysts such as zeolites with an extended system of narrow pores, but the characteristic clefts containing the enzyme active centers are relatively short, and accessibility to the active center is rarely a problem in comparison. However, confusion has entered certain areas even of homogeneous catalysis from a failure to be aware of diffusion limitations. Selectivity is a common factor in all three types of catalysis, although obviously most apparent in enzymes; and it may be that all three branches can learn a great deal from each other about the relative importance of electronic and steric factors in particular types of reactions, with possible advantage in the long run to chemical industry. From enzymes we learn that the active site is associated with several binding groups in an essentially flexible structure based upon a number of specific conformations in the presence or absence of substrate or other molecules; the possibility of producing molecular catalysts of this type for new reactions would certainly be of great importance in industry. Here the experience of organic and inorganic chemists in the synthesis of models for existing enzymes should be of the utmost value. The cooperation of enzymologists in designing models, of organic chemists in synthesis and of workers in homogeneous catalysis doing mechanistic studies, is strongly urged. In the area of heterogeneous catalysis by metal crystallites there is an overwhelming problem of the interaction between neighbouring metal sites, and the study of catalysis by metal cluster compounds in solution (or even by simple dimers) offers very definite help in this area.

The role of water, both structured and unstructured, in catalysis generally is largely obscure, and its study by modern techniques, such as NMR, the various kinds of pulsed NMR, dielectric-loss and microwave methods, etc., forms a common area of effort, e.g. as between proteins and zeolites at the present time. Ther may be other cases where elaborate physical techniques which are available and used in one area of catalysis should be brought to bear in other areas. Studies of all kinds of molecular motions, of energetic and configurational fluctuations and their relation to the activation of adsorbate to the transition state and to the microscopic course of the reaction are of common interest to all three fields of catalysis.

In nature, reactions are coupled by specific enzyme systems in such a way that the negative free-energy change in one reaction is, as it were, used to drive a subsequent synthesis with a positive change in free energy. Here of course the role of the mitochondrion in linking the oxidation of NADH by gaseous oxygen (ΔG negative) to the formation of three molecules of ATP from ADP (ΔG positive) is the classical example. The analogies in industrial and laboratory chemistry in fact involve the isolation of products (for which ΔG is negative), their subsequent reaction with the input of thermal, electrical or light energy to facilitate a cyclic reaction to give the original catalyst and the final product (which has ΔG positive with respect to the original reactants). It is suggested that the further study of these coupled reactions of biochemistry, with their highly specific catalysts, should one day bear fruit in industrial synthesis.

There is also the principle of induced enzyme synthesis in which the introduction of substrates into the cell may cause the synthesis of specific enzymes for their own reaction. It may be that analogies are possible in heterogeneous catalysis if it were feasible to construct a surface sufficiently flexible to adapt itself from one substrate to another, if necessary under the influence of a modifying reagent.

The theories of catalysis, which will be theories of the free energy of activated complexes in relation to reactants and products, must ultimately depend on solutions of the time-dependent Schrödinger equation. In the meantime, all three branches are developing their own intuitive theories, rules and nomenclature, and we consider that only good can come from general awar ness of these theories and their supporting evidence by all three groups of workers.

In summary, six key problem areas were defined which apply equally to all three fields. These were: characterization of reaction intermediates; mod of activation of bound (chemisorbed) molecules to products; methods of esti mating concentrations of active sites for the purpose of expressing kinetic data as turn-over numbers; development and exploitation of novel new kineti methods (including coupling of spectroscopic tools with kinetic measurement isolating factors which lead to selectivity in product formation; and furth extensive use of tracers in developing reaction schemes and isolating rate-limiting steps (where possible) through the use of isotope effects. In addition the establishment of a catalyst bank from which samples of standard catalysts and purified enzymes can be withdrawn for comparative purposes is recommended.

HYDROGENATION/DEHYDROGENATION

CHAIRMAN: G. C. Bond
Department of Industrial Chemistry, Brunel University, Kingston Lane, Uxbridge, Middlesex, England

RECORDER: B. R. James
Department of Chemistry, University of British Columbia, Vancouver 8, B.C., Canada

MEMBERS: R. L. Burwell, Jr. J. E. Coleman
P. Cossee C. Franconi R. J. Kokes
L. Vaska J. J. Villafranca

1. PREAMBLE

The scenario for the future suggests that catalytic hydrogenation and related processes will play a vital role in energy supply and distribution. Forecasts indicate that hydrocarbon energy sources will be seriously depleted before nuclear energy has taken over entirely. Coal, shale, and tar sands supplies may be necessary in between in order to supply the needed energy. Catalytic processes will come into the picture at three points:

(a) When coal becomes the dominant energy source it will probably be converted (in part) to liquid or gaseous fuels by catalytic processes so that it is in a convenient form for use in all modes of transportation.

(b) When nuclear reactors become the dominant energy source they will probably be located away from population centers. It has been shown that if such reactors are far enough away from the user (e.g. more than 500 km) it will be cheaper to electrolyse water to hydrogen and transport this by pipelines than to transport electricity by power lines. Thus hydrogen may not only be used for fuel in fixed and mobile units but may be reconverted to electricity by the user. Such reconversion will require fuel cells which in turn require efficient electrodes composed of catalytic materials capable of activating hydrogen (and oxygen).

(c) Again, nuclear reactors will probably never be used in (*small*) energy sources for transportation. Liquid or gaseous chemical fuels will still be the most convenient form. Hydrogen may be used directly or converted to other fuels. These other fuels will probably be made by catalytic processes utilizing cheap hydrogen produced by nuclear reactors. Furthermore the wide availability of hydrogen will make it economically attractive to utilize hydrogenation reactions of all kinds.

The above emphasizes very strongly that adequate planning for the future requires development of new, and refinement of old, catalytic processes involving hydrogen. Unless we are prepared for this, we may, for some time, be faced with energy starvation.

2. SURVEY OF CLASSES OF HYDROGENATION/DEHYDROGENATION REACTIONS

A survey of the types of reactions involving hydrogenation-dehydrogenation, irrespective of the nature of the catalyst (homogeneous, heterogeneous, enzymatic), indicates a possible classification as follows:

A. Hydrogen Activation, and Reactions of Hydrogen in the Absence of Added Substrates

(i) Adsorption/coordination of H_2 (heterolytic, homolytic).

(ii) *Ortho-para*-hydrogen equilibration.

(iii) Hydrogen isotope exchange.

(iv) Hydrogen atomization and atom recombination.

(v) Hydrogen-ion discharge and ionization of H_2.

All these processes have been studied on surfaces; homogeneous and enzym[e] systems have been recognized for (i) - (iii); reaction of metal ions in sol[u]tion with hydrogen atoms has been studied to a certain extent.

B. Hydrogenation of Unsaturated Functions (Including Use of Hydrogen Isotopes)

(i) Carbon-carbon multiple bonds (C=C, C≡C, C=C-C=C, aromatic rings, etc.). Homogeneous and heterogeneous catalysts are known for all these processes; enzymatic hydrogen-transfer processes are involve[d] at certain carbon-carbon double and triple bonds.

(ii) Carbon-oxygen multiple bonds (C≡O, CO_2, >CO, -CHO, $-CO_2H$, α, β-unsaturated aldehydes, etc.). Such reductions have been accomplishe[d] using heterogeneous catalysts; homogeneous systems are restricted to the reduction of ketones and aldehydes, while enzymatic systems reduce carbonyl and carboxylic functions to alcohols.

(iii) Carbon-nitrogen multiple bonds (-C≡N, >C=N-, pyridine, etc.). The[se] have been reduced with only heterogeneous and homogeneous catalyst[s].

(iv) Carbon-sulfur multiple bonds (>C=S, thiophene, etc.). Homogeneous and heterogeneous catalysts have been used.

(v) Multiple bonds not involving carbon ($-NO_2$, -NO, -N=N-, NO_2, NO, SO[2], SO_3, etc.). Heterogeneous systems can effect all such reductions to a certain extent; homogeneous catalysts have been reported for the first three types, while enzyme systems are involved in the reduction of -N=N- systems.

C. Dehydrogenation

These can be classified as the reverse of those reactions listed in B(i) - (iv), plus a further classification: (v) Oxidative dehydrogenation, wherein abstracted H_2 reacts with O_2 or other hydrogen acceptors to shift the dehydrogenation equilibrium. Heterogeneous catalysts have been used for all these types of dehydrogenation; homogeneous studies are few and have involve[d] formation of carbon-carbon double bonds from saturated systems. Classification (v) incorporates certain steps in biological oxidation systems.

D. Hydrogen Transfer (Coupled Hydrogenation—Dehydrogenation)

(i) Biological (e.g. NAD-NADH, flavin systems).

(ii) Heterogeneous (e.g. isomerization via carbonium intermediates).

(iii) Homogeneous (e.g. cyclohexadiene → cyclohexene and benzene).

E. Reductive Hydrogenation of Inorganic Compounds

(i) Homogeneous reduction of cations and anions in solution (homogeneous, enzymatic).

(ii) Reduction of solid and molten oxides, sulfides, etc. (this excludes *oxidative* addition of H_2 to complexes).

(iii) Reduction of O_2 to H_2O_2 and of N_2 to NH_3 (heterogeneous, enzymatic).

F. Hydrogenolysis

(i) Splitting of metal-metal and metal-carbon bonds (homogeneous and heterogeneous).

(ii) Splitting of carbon-carbon single bonds, removal of carbon deposits and surface carbides (heterogeneous).

(iii) Splitting of carbon-X single bonds, e.g. C-Cl, C-O, C-N, C-S, etc. (mainly heterogeneous but homogeneous reductions of carbon-halogen bonds are known).

(iv) Splitting of X-Y bonds (e.g. Cl-Cl, =N-N=, -N=N-, etc.).

(v) Coal hydrogenation.

G. Other Reactions

(i) Skeletal isomerization of hydrocarbons (heterogeneous).

(ii) Double bond and geometrical isomerization (homogeneous, heterogeneous, enzymatic).

(iii) Isotope-exchange reactions (homogeneous, heterogeneous, enzymatic).

3. CLASSIFICATION OF CATALYTIC SYSTEMS AND FUTURE DEVELOPMENTS

We feel it useful to have an overview of the types of catalyst systems capable of hydrogenation/dehydrogenation catalysis.

A. Homogeneous Systems

(i) Complexes in true solution, including charge-transfer complexes, phthalocyanines, boranes, and other non-metallic systems.

(ii) Supported solutions of complexes.

B. Heterogeneous Systems

(i) Conventional solids (metals, alloys, oxides, sulfides, etc.).

(ii) Supported complexes and complexes in the pure solid state.

(iii) Molecular complexes in the solid state (porphyrins, charge-transfer complexes, etc.).

(iv) Electrocatalysts.

C. Enzymatic Systems

(i) Enzymes *in vivo*.

(ii) Enzymes *in vitro*.

(iii) Isolated prosthetic group.

(iv) Enzymes supported on solid porous matrices.

We wish to press for a more broadly ranging search for new and improved hydrogenation/dehydrogenation catalysts. While it is true that a feature catalysis research is the continued striving for improvement in catalyst quality, we believe that true progress will depend on the discovery of qui novel systems, and that this will require the imaginative and creative ene ies of chemists of all kinds.

We note for example that some metals and metal oxides have not been stud ied in depth; that metal sulfides have only been sparsely examined; and we have before us continually as an ideal the activity and specificity of enzymes.

4. MAIN RECOMMENDATIONS

After long and tortuous discussion, we have selected five fields as bein those in which sustained research effort would be most rewarding, either scientifically or technologically or both. This selection has been based the criterion of interaction between the three types of catalysis. We hav not made any recommendations of substance which relate solely to one partic ular field. We have felt unable to make any proposals relating to electrocatalysis, while recognizing that this may yet turn out to be an area of great importance: we note for example continuing interest in fuel cells ar in electrocatalytic synthesis.

Most of the proposals relate directly to hydrogenation/dehydrogenation catalysis. However, since kinetic and mechanistic analysis in this area ha a long and honorable history, and since many problems remain, we have submi ted some ideas on the value of instrumental techniques which might be espec ially pertinent. We have made no specific recommendations concerning dehydrogenation since this is chiefly the province of heterogeneous catalysis: this area also impinges on 'Catalytic Activation of Saturated Hydrocarbons'

A. Hydrogen Activation and Subsequent Steps

Hydrogen activation implies reaction of the molecule with a substance to form a species more reactive than the original molecule. Usually such acti vation involves rupture of the hydrogen-hydrogen bond but in some cases activation may involve merely close interaction of the hydrogen molecule wi the bonding substances. Reaction of hydrogen to form such activated specie can be very rapid; in fact with some metal surfaces nearly every collision results in dissociation. It is very difficult to predict *ab initio* which substances will activate hydrogen and which substances will not. For examp if we examine similar metals such as Ni and Cu, we find Ni efficient as an activator and Cu inefficient. Such comparisons can also be made for other similar substances. Since hydrogen activation is the *sine qua non* for hydr genation reactions, this lack of understanding is serious. Better understa ing of this process would enhance the probability of effective design of ne catalyst systems.

It is widely believed that with homogeneous systems the activation step is well understood. This is not so for heterogeneous systems. Phenomena such as site-to-site migration between widely separated active sites are still described in terms of speculative mechanisms. Spillover, wherein hydrogen is dissociated on one site and migration occurs across the surface, or from one particle to another, to react, requires further study.

Measurements of kinetic and thermodynamic isotope effects can provide useful information on the molecular parameters that define the activated hydrogen and the transition state. Measurements of the *ortho-para* conversion (a form of activation) coupled with hydrogen-deuterium exchange-rate studies provide information on the relative site-to-site mobility of hydrogen or the paramagnetism of the activating sites. Such studies, especially if a sound theory for *para*conversion of bound molecular species can be developed, will do much to amplify our knowledge of the factors that govern the stability and reactivity patterns that define hydrogen activation.

In homogeneous catalysis the current pressing question seems to be what electronic and structural factors govern the reactivity of the dissociatively bound hydrogen with the other substrate. There are systems wherein the hydrogen is dissociated but will not undergo further reactions with any unsaturated species. Similar examples are found (but not as often) in heterogeneous systems; for example, magnesium oxide will catalyse H_2-D_2 exchange at low temperatures, but does not catalyse ethylene hydrogenation. It seems critical that we understand better the factors that control subsequent reactivity of the substrate with activated hydrogen. Indeed, this is a significant part of the overall problem of hydrogen activation and for homogeneous catalysts appears to be the most critical factor.

B. Utilization of Hydrogen by Enzymes

Future research in the areas of hydrogenation/dehydrogenation catalysis should involve the search for biochemical systems which can incorporate hydrogen. The two areas of homogeneous and heterogeneous catalysis already use gaseous hydrogen for hydrogenation, but there are few if any analogies in biochemical systems. Biochemical hydrogenations occur by the use of pyridine nucleotides and flavin-requiring enzymes, and can be thought of as occurring by a hydride and proton transfer or by a two-electron and two-proton transfer. The exact mechanism is not precisely known. The study of hydrogen-fixing bacteria may therefore allow the isolation of the enzymatic system which carries out this process. All the principles found from heterogeneous and homogeneous catalysis involving hydrogen will be essential in this study. The enzymatic system will then be available for study by the multitude of spectroscopic and X-ray techniques already used to study the structure and function of enzymes. These principles should then be of great use to scientists in the areas of homogeneous and heterogeneous catalysis in understanding further the basic nature of hydrogen transfer. Together with the fixation of hydrogen, the transfer of hydrogen to specific molecules should be studied with a view to hydrogenating them stereoselectively in cases where it is difficult or impossible to do so homogeneously or heterogeneously. Synthetic methods for optically active compounds would be especially useful. Therefore the specificity of enzymatic reactions involving hydrogenation by hydrogen could be utilized for industrial processes. The search for metal hydrides in biological systems could be a by-product of this search, which could also help advance knowledge in the area of hydrogenation via metal hydrides.

C. Hydrogenation of Small Inorganic Molecules

Attractive problems in the area of hydrogenation/dehydrogenation catalysis are the hydrogenation of small inorganic molecules such as CO, CO_2, and NO/NO_2.

The direct reduction of CO_2 (other than to CO by the water gas shift) is difficult by heterogeneous methods, and has not been achieved so far in homogeneous systems. Considering the long-term future when nuclear power will have become the predominant power source, hydrogen from electrolysis of water might be reacted with atmospheric CO_2 as the carbon source in the productic of a liquid fuel such as methanol or organic molecules of greater complexit The scope would be enormously wide, and the problem deserves attack by heterogeneous, homogeneous and in particular by enzymatic catalysis.

The heterogeneous hydrogenation of CO to methanol is a well-established process. Since methanol may become an important liquid fuel in the future, further attempts to improve the existing process or to find new approaches are highly recommended. The formation from CO and H_2 of higher-molecular-weight compounds has been accomplished by the Fischer-Tropsch synthesis; it is, however, a neglected area of research. Renewed attention to this type of process may become highly desirable in the near future.

The possibility of using enzymes for converting CO to industrially useful compounds should not be discounted simply because of the known sensitivity of certain enzymes to this molecule. The aim here would again be the production of liquid, easily transportable and harmless (e.g. sulfur-free) fue and industrial solvents.

The reduction of NO and NO_2 deserves the attention of all three types of catalysis. In addition to seeking new and more effective ways of removing these gases as atmospheric contaminants, it might be possible to utilize th products of their reduction, as for example in making hydrazine or hydroxylamine.

D. Stereoselective Homogeneous and Heterogeneous Catalytic Synthesis

Further effort should be made by workers in the areas of homogeneous and heterogeneous hydrogenation catalysis to produce catalysts that emulate the stereoselectivity exhibited by enzyme systems. Considerable progress has been made recently in the homogeneous area by using metal complexes containing chiral ligands, and in the heterogeneous area using metal catalysts modi fied by adsorption of such optically active species as amino acids

Hopefully one could prepare homogeneous catalysts containing appropriately chosen ligands which would give even higher optical yields than those so far developed; and one would hope to cover properly chosen metal particles, so as to leave 'optically active' holes. Of course, further advances in these areas would presumably cause both types of catalysts to become restricted in the types of compound they would hydrogenate.

Asymmetric hydroformylation as well as hydrogenation should receive furthe attention; for example, the production of optically active aldehydes and alc hols from olefins in high optical purity. Encouragement should be given to studies in stereospecific hydrogenation, for example, the *cis*-addition of hydrogen, deuterium or tritium to unsaturated centres in more complex organi molecules such as steroids, terpenes, and other natural products, which appears to be becoming a powerful tool for the synthetic organic chemist.

E. Methods of Investigation and New Techniques

A number of spectroscopic and magnetic techniques including ESR and NMR have already been successfully applied to obtain useful information on active sites, structural and electronic properties of solids and surfaces, and on adsorbed species.

The X-ray technique has done entirely different things for enzyme and homogeneous catalysis as compared to heterogeneous catalysis. Whereas it has given detailed information not only on enzyme structures but also on the enzyme-substrate combination and the structure of intermediates in homogeneous catalysis, it has failed to give information on active sites and adsorbed species in heterogeneous systems. It has, though, been very helpful in the characterization of structure and texture of the bulk of solid catalysts.

High-resolution NMR has given very valuable kinetic and structural information in homogeneous and enzymatic catalysts.

New sophisticated resonance techniques have been developed, such as all forms of double resonance (nuclear-nuclear, Overhauser effects and electron-nuclear double resonance), pulsed techniques, and Fourier transforms, which have not yet been extensively used.

A third type of new and sophisticated technique is emerging in the form of electron diffraction and electron spectroscopy (low-energy electron diffraction, ESCA, and Auger spectroscopy).

Without being specific, a number of major problems can be identified:

(1) The non-generality of many of the techniques because of their interim characteristics. For instance, all diffraction techniques require a high degree of ordering of sufficient size; resonance techniques have their limitations in line broadening because of dipole interactions and relaxation effects, etc..

(2) Many techniques have difficulty in separating bulk and surface effects and in distinguishing between active sites or species and their precursors. Very often only the latter are observed because of their much higher concentration.

(3) Many techniques are time-consuming, in particular the new electron spectroscopy methods. They require repeated rapid scanning and integration of the results over long periods.

(4) Many of the techniques can be applied only under 'clean' conditions: ultra-high vacuum, room temperature, normal pressure or in the case of resonance and double-resonance techniques only at the temperatures of liquid H_2 or even liquid He.

In order to further our knowledge of catalysis, we believe it is most important to be able to study our systems while catalysis is proceeding. We regret that physical studies have been so largely confined to examining adsorbed or coordinating molecules and we applaud that little done on reacting systems. We also note that the majority of work to date has been of necessity performed with very clean materials and often with single crystals. It is highly desirable to develop comparable methods of examination applicable to technical catalysts.

In view of the above it is therefore recommended that special emphasis be placed on those technical developments that will make it possible in the future:

I. To observe our systems under actual reaction conditions.

II. To distinguish better between bulk and surface properties.

III. To adapt the timescale of the technique to detect short-lived species.

5. CONCLUDING REMARKS

Catalysis by enzymatic, homogeneous, and heterogeneous systems is similar insofar as each system makes processes possible that are otherwise (practically) impossible. Each of these systems, however, has its own individual characteristics. For example, enzyme catalysis invariably involves ions ir aqueous solution in a limited temperature range, whereas heterogeneous cata lysis more often involves gases and is operable over a wide temperature ran As individuals, these systems (and the scientists who study them) have thei own culture. In the foregoing discussion we have tried by enforced inter- course to promote cross fertilization of these somewhat parochial cultures. The results can be viewed as a series of challenges: Can heterogeneous cat lysts provide the stereospecificity of enzymes? Can homogeneous catalysts effect the same reactions as heterogeneous catalysts? Can enzymes activate molecular hydrogen as both homogeneous and heterogeneous catalysts do? The questions are phrased not to provide mindless challenges; we do not have to hydrogenate carbon monoxide 'because it is there'. These challenges are is sued because we must be clever if we are to meet the energy deprivations th are in sight. And reaction to these challenges may provide the means to co with our collective future.

There is danger also in the recommendation we have made. This is espe- cially so if workers in an area we have not included should have to curtail their fruitful work and follow our recommendations. Prognostication on the basis of *now* may overlook the reaction of the future in areas that *now* seem limited in scope. All three disciplines need, in particular, a means of securing information on the step-by-step atomic motions that constitute rea tion. Physical techniques interfaced to the live catalyst system are neede and different systems have different needs. There is, however, a common ob- jective. Because of this, cross-fertilization by a forced mixture of the three groups should be continued.

LITERATURE RECOMMENDED

Homogeneous Catalysis

COFFEY, R.S. (1970). Recent advances in homogeneous hydrogenation of carbon-carbon multiple bonds. In *Aspects of Homogeneous Catalysis, Vol. 1*, (ed. Ugo, R.), (Carlo Manfredi, Milan), p. 3.

KWIATEK, J. (1971). Hydrogenation and dehydrogenation. In *Transition Metals in Homogeneous Catalysis*, (ed. Schrauzer, G.N.), (Marcel Dekker, New York), p. 13.

JAMES, B.R. (1973). *Homogeneous Hydrogenation*, (Wiley, New York).

Heterogeneous Catalysis

BURWELL, R.L., TAYLOR, K.C., READ, J.F. and HALLER, G.A. (1969). *Adv. Catalysis*, 20, 1.

BOND, G.C. (1962). *Catalysis by Metals*, (Academic Press, London and New York

Enzyme Catalysis

BOYER, P. (1970-1972). *The Enzymes, Vols 1-8*, (Academic Press, London and New York).

NITROGEN FIXATION

CHAIRMAN: J. Chatt
The Chemical Laboratory, University of Sussex, Falmer, Brighton BN1 9QJ, England

RECORDER: M. Boudart
Department of Chemical Engineering, Stanford University, Stanford, California 94305, USA

MEMBERS: E. Bayer W. Beck H. H. Brintzinger
J. A. Ibers K. Jonas H. L. Krauss
J.-C. Marchon C. Veeger

Ammonia synthesis from dinitrogen and dihydrogen has been carried out catalytically since 1913. More recently, great progress has been accomplished in the understanding of nitrogen fixation with nitrogenase systems isolated from anaerobic organisms. Recently, efforts have been made to reduce catalytically dinitrogen complexes in solution. In this report, the state of knowledge in these three areas of heterogeneous, metalloenzyme and homogeneous catalysis will be assessed with the aim of delineating promising avenues of research.

HETEROGENEOUS CATALYSIS

Introduction

The metals of the iron group are by far the most active heterogeneous catalysts. The activity increases in the order Fe < Ru < Os, but the last two metals were thought until recently to be too expensive. Thus, the catalyst has hardly changed since the days of Haber and Bosch. It consists of metallic iron particles about 25 nm in diameter containing small amounts (c. 3%) of unreduced oxides called promoters, typically Al_2O_3, K_2O, and CaO. The cost of the catalyst is very small: about \$ 0.1 per ton of ammonia (\$ 20). In spite of the unlikely prospect of a significant lowering of catalyst cost, active work in ammonia synthesis is still carried out in the USSR, Japan, Italy, and the USA. Extensive work in Holland and West Germany has been discontinued. Interest in ammonia synthesis appears largely due to the expectation that general concepts and tools of importance to the entire catalytic community will fall out from work with this model reaction. This has happened many times spectacularly in the past. Besides, there always looms the prospect of such a drastic improvement in catalyst activity that significant savings in plant investment might become a reality as a result of a drop in process temperature and pressure.

What is Known

With respect to the commercial catalyst, the role of the first promoter, Al_2O_3, is generally agreed upon to be only textural. This means that Al_2O_3 does not change the catalytic activity of the Fe but only preserves its porous texture. The mode of action of the alkali and alkaline earth promoters is still a matter of controversy.

With respect to reaction mechanisms, the consensus of opinion is that the chemisorption of dinitrogen on the Fe surface is the rate-determining process and that, at least above 700 K, the most abundant surface intermediate contains only nitrogen. The unusually good and extensive kinetic data can best be fitted on the assumption of a broad non-uniformity of surface sites. The latter is also revealed by many kinetic and thermodyamic studies of chemisorption.

It must be noted that dinitrogen forms unreactive complexes at metal surfaces such as Ni or with coordinatively unsaturated ions on oxide surfaces. These are of interest in relation to similar complexes studied in solution (see Section on Homogeneous Catalysis).

Some Problems

The nature of the most abundant intermediate is not known: is it mononitrogen or is it dinitrogen? In the latter case, is the adsorption terminal (end-on) or bridged on two or more atoms? These questions are of general interest in all areas of the chemical reactivity of dinitrogen.

Where is this intermediate chemisorbed? Everywhere on the iron surface or at privileged sites, perhaps on (111) planes? An answer would contribute significantly to the central problem in heterogeneous catalysis of the structure sensitivity to certain reactions.

A related aspect concerns the nature of the non-uniformity of the surface. If only sites on (111) planes are active, why does everything happen as if they were a collection of non-uniform entities? What is the nature of this non-uniformity? If it is due to interactions between the nitrogen and the surface, are these interactions responsible for the beneficial action of potassium? It must be noted that this idea apparently led Ozaki *et al.* to the first significantly improved catalyst reported in many years. It consists of Ru metal supported on carbon and activated by K metal. This recent Japanese discovery, even if it is not developed commercially, illustrates the potential impact of these general concepts on catalyst research, though the concepts are still only speculative, and are not generally accepted by workers in the field.

Finally, the mode of action of the first textural promoter, Al_2O_3, deserves further examination. Is Al_2O_3 acting like a protective patchy skin at the surface of the metallic particles to prevent their welding together (sintering)?

METALLOENZYME CATALYSIS

Introduction

Ammonia is the result of nitrogen fixation by symbiotic and free-living bacteria. Among the latter are *Klebsiella pneumonia* (*K*), *Clostridium pasteurianum* (*C*) and *Azotobacter vinelandii* (*A*). The enzyme nitrogenase has been obtained from these three species in a state of high purity. It can be separated into two proteins called, by some, the Mo-Fe and the Fe protein. These may be broken down into sub-units.

In 1960, Carnahan *et al.* discovered that nitrogenase was active in the fixation of nitrogen only in the absence of dioxygen and in the presence of a source of adenosine triphosphate (ATP), an electron carrier — ferredoxin, and a reducing agent — sodium pyruvate. This discovery sparked a large number of investigations in the UK, the USA and Holland. One of the most important by-products of this work has been the finding that nitrogenase is not as specific as most enzymes. Indeed, it also reduces readily a number of dinitrogen analogues such as C_2H_2, N_3^-, N_2O and CN^-.

What is Known

The extreme sensitivity of nitrogenase toward dioxygen applies equally well to the enzyme extracted from aerobic and anaerobic bacteria. Thus nitrogen fixation by aerobic *Azotobacter* is strictly anaerobic.

The anaerobic pathway, as presently understood mainly from ESR studies, can be represented by the following cycle where all terms have been previously introduced except ADP which denotes adenosine diphosphate.

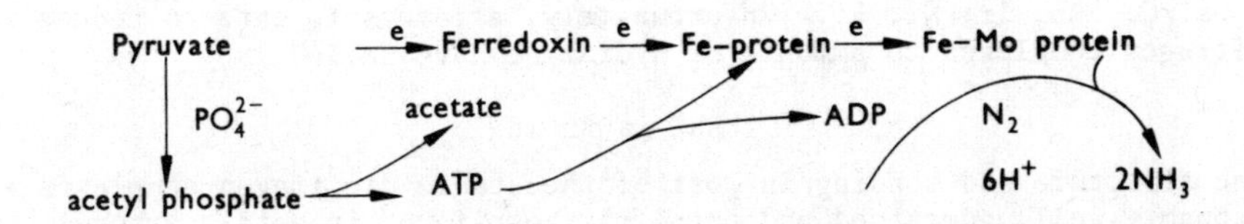

Magnesium ion, not included in the above cycle, is also essential. The characterization of nitrogenase available today is summarized in the following Table:

Protein	Species	Mol. wt.	Number of atoms per molecule			Number of subunits
			Fe	Labile S	Mo	
Mo-Fe	*K*	220 000	17	17	1-2	4 non-identical
	C	220 000	22	22-24	2	4 non-identical
	A	270 000	33	26	2	?
Fe	*K*	68 000	4	4	none	2 identical
	C	55 000	4	4	none	2

Some Problems

Unknown remain the nature of the subunits, the reason for the sensitivity to dioxygen, the oxidation states and role in dinitrogen bonding to the Fe and Mo sites, and the occurrence of a metal hydride suggested by isotopic exchange between water and dideuterium. The nature of the ESR active group ings of both the Mo-Fe protein and the Fe protein remains to be resolved, s that the oxidation states of Fe and Mo and their ligand fields become known together with the number of reducing equivalents transferred. Besides, the sequence analysis and X-ray diffraction examination of the proteins remain to be done. Finally, what are the interactions between the two proteins, t role of ATP and Mg^{2+}, and the exact point at which ATP is hydrolysed?

HOMOGENEOUS CATALYSIS

Introduction

Following the preparation in 1965 by Allen and Senoff of the first dinitr gen complex, $[(Ru(NH_3)_5(N_2))]^{2+}$, a great number of stable metal complexes containing dinitrogen in terminal or bridging position have been characterized. Among the transition metals involved are Ti; Cr, Mo, W; Mn, Re; Fe, Ru, Os; Co, Rh, Ir; Ni, Pt. Unfortunately, attempts to date to reduce stab dinitrogen complexes to ammonia or hydrazine have failed.

What is Known

The structure and bonding in most of the stable dinitrogen complexes are reasonably well understood and parallel those found in metal carbonyl complexes. Empirically, a number of systems have been reported to convert dinitrogen to ammonia or hydrazine but it must be stressed that none of these *homogeneous* systems effects this conversion in catalytic fashion. However, catalytic reduction of dinitrogen has been reported in some related aqueous *heterogeneous* systems. Of special interest is the observation that organic nitrogen-containing compounds can be obtained from dinitrogen under mild conditions.

Some Problems

What are the specific differences between those unstable dinitrogen complexes where the dinitrogen ligand can be reduced and those stable complexes where it cannot be reduced? What is the mechanism of reduction? What is the role of hydride, proton or electron transfer in the reduction?

GENERAL RECOMMENDATIONS

We believe that the study of nitrogen fixation is important for industrial and nutritional purposes. It could lead to new processes for the direct production from dinitrogen of ammonia or nitrogen-containing organic chemicals. In agriculture, mutated or genetically altered nitrogen-fixing bacteria coul contribute to increased ammonia production in the field in a biologically an ecologically balanced manner. For these reasons, we recommend:

(1) The study of nitrogen fixation by heterogeneous, homogeneous, and metalloenzyme catalysis.

(2) The determination of those properties of metals and metal complexes that allow (a) the mere addition of dinitrogen, (b) its addition followed by its reduction by any means but (c) more specifically by hydrogen, and (d) in addition in a catalytic way.

(3) The clarification of the mechanisms in known N_2 reduction processes by the identification and characterization of reactive intermediates in all three fields of catalysis.

(4) The study of the catalytic reduction of dinitrogen-like compounds, especially CN^-, CO and NO, which may lead to more effective catalysts for the removal of poisonous substances from the environment.

(5) The solving of the problem as to whether the activation of dinitrogen needs a single or multiple metal site.

(6) The search for systems in which the dinitrogen ligand can be reduced electrochemically or photochemically.

SPECIFIC RECOMMENDATIONS

Heterogeneous Catalysis

We recommend the urgent exploitation of new spectroscopic and physical tools to examine surface composition, structure and bonding. Examples of studies are the determination of the distribution of promoters at the surface of conventional Haber catalysts and the examination of intermediates at the surface of working catalysts. The time seems ripe for a systematic study of nitrogen chemistry on clean metal surfaces by field-emission microscopy, field-ion microscopy, low-energy electron diffraction, flash desorption, work function, etc., to establish the coordination chemistry at surfaces. We also recommend the preparation and study of a new generation of supported metal catalysts containing clusters of zerovalent metals or individual metal ions, in order to approach the possible behaviour of multinuclear metal complexes in solution or immobilized on supports. Finally, we suggest a fresh attack on the problem of surface non-uniformity for ammonia synthesis at metal surfaces, as this phenomenon may well ultimately prove to be the only significant difference between chemistry at solid surfaces on the one hand and chemistry with coordination complexes and metalloenzymes on the other hand. The activation of dinitrogen seems to be a uniquely favorable subject for such a unifying investigation.

Metalloenzyme Catalysis

First, we wish to bring to attention the fact that, in the future, mutants may be obtained with a higher capacity of nitrogen fixation and less sensitivity towards dioxygen. Secondly, it must be considered that other yet unknown biological systems for nitrogen fixation may be discovered. For instance, in certain areas of New Guinea, the population can live on a very low protein diet provided that a certain type of potato is included. It seems possible that a certain human intestinal flora is able to synthesize essential amino acids from dinitrogen.

Homogeneous Catalysis

We recommend urgently the study of electronic and geometric requirements of co-ligands which would permit the reduction of the dinitrogen ligand. Some of these co-ligands might resemble those found in enzyme systems. Further, we recommend preparation and study of polynuclear dinitrogen complexes with oligomeric and polymeric co-ligands. Finally, the chemistry of complexes which contain possible intermediates in the reduction of dinitrogen, such as diimine and hydrazine complexes, should be studied.

BIBLIOGRAPHY

THOMAS, J.M. and THOMAS, W.J. (1967). *Introduction to The Principles of Heterogeneous Catalysis*, (Academic Press, London and New York).

POSTGATE, J.R. (ed.) (1971). *The Chemistry and Biochemistry of Nitrogen Fixation*, (Plenum Press, London).

CHATT, J. and LEIGH, G.J. (1972). Nitrogen fixation, *Chem. Soc. Rev.*, 1, 121, (for recent reviews in the enzyme and homogeneous areas).

OXYGEN ACTIVATION

CHAIRMAN: C. R. Jefcoate
Department of Biochemistry, University of Wisconsin, Madison, Wisconsin 53706, USA

RECORDER: V. Ullrich
Physiologisch-Chemisches Institut der Universität des Saarlandes, 665 Hamburg, Germany

MEMBERS: E. Antonini A. Cimino J. E. Germain
R. D. Gillard J. A. Ibers R. Lontie
A. Martell R. Ugo

INTRODUCTION

Molecular oxygen or, as it is correctly termed, dioxygen, is a potentially strong oxidizing agent with regard to the thermodynamics of the change to water. However, this change must generally occur in a sequence of one-electron steps and is kinetically unfavourable, in part owing to the adverse potential for the first electron transfer. A second factor which has been implicated is the restriction imposed by the requirement for 'spin conservation' in the overall electron transfer to the *triplet* dioxygen. Thus the dioxygen molecule is a metastable species in the presence of organic compounds, which is the essential condition in maintaining life in an atmosphere of oxygen.

In many chemical and biological processes, however, dioxygen is required for the oxidation or oxygenation of organic compounds. In these cases the dioxygen requires an activation process to carry out these reactions.

Catalytic processes which lead to activated species of oxygen and dioxygen are known in the fields of homogeneous, heterogeneous, and enzyme catalysis. Owing to the complex chemistry of oxygen in its various oxidation states and the high reactivity of most of the active oxygen and dioxygen species, little is known about the mechanisms involved in the oxygen-activation reactions.

There are various ways to change a stable electronic configuration of the oxygen molecule. The most common activation processes involve electron transfer to molecular oxygen by metal ions in solution, on surfaces or in enzymes. Although the conditions used in the three fields of catalysis are widely different, it is possible that certain common mechanisms are involved. A selection of reactions that involve oxygen activation in homogeneous, heterogeneous, and enzyme catalysis is listed in Table I. We have concentrated on metal ion catalysis, since this provides common ground between the three disciplines. However, in many enzymatic processes, dioxygen is activated without the participation of a metal ion, by protein-bound organic molecules which can undergo oxidation-reduction changes.

Compared to the chemical reactions in solution or at heterogeneous catalyst surfaces, enzyme reactions are very specific with regard to products formed, and further reaction is generally avoided. There is frequently a basic difficulty in comparing heterogeneous oxidation processes involving dioxygen with reactions which occur in solution since the primary step which

TABLE I

Examples of Oxygen-Activation Reactions

Substrate	Product	Catalyst	Conditions
	1. Oxygen addition		
CO	CO_2	Many oxides and metals (HET)†	25 °C
Ethylene	Ethylene oxide	Ag (HET)	200 °C
Benzene	Phenols	Fe^{2+}, EDTA, H_2O_2 (HOM)	30 °C pH 5-8
Hydrocarbons	Alcohols	Many hydroxylases (Fe, Fe-heme) (ENZ)	Physiological pH and temp.
Benzene	Maleic anhydride	V_2O_5-MoO_3 (HET)	450 °C
Catechol	COOH COOH	Catechol dioxygenase (ENZ) (Fe)	37 °C pH 7.5
$P(Ph)_3$	$OP(Ph)_3$	$Pt(P(Ph)_3)_n$ (HOM)	25 °C benzene
	2. Oxidation without insertion of oxygen from dioxygen		
Propylene	Acrolein	Bi-Mo oxides (HET)	450 °C
Butene	Butadiene	Bi-Mo oxides (HET)	450 °C
Propylene	Benzene	Bi-P oxides (HET)	400 °C
Ethylene and benzene	Styrene + H_2O	Pd(II)	25 °C aqueous acetic acid
Diphenols	Quinones	Cu-oxidases Laccase (ENZ)	37 °C pH 7.5
R-CH_2CH_2-R' (stearyl CoA)	R-CH=CHR' (oleyl CoA)	Fatty acid desaturase (possibly Fe-hemecenter) (ENZ)	37 °C pH 7.5

† (HET) signifies heterogeneous; (HOM), homogeneous; (ENZ), enzymatic.

determines selectivity on surfaces, particularly at high temperatures, may involve a change only in the other reactant. This sequence of events is also realized in some reactions in solution. For example, in the enzymatic oxidation of ascorbic acid by ascorbate oxidase and molecular oxygen, ascorbic acid is initially oxidized sequentially by enzyme-bound cupric ions whic are *subsequently* rexodized by molecular oxygen. In such cases the controlling process is substrate activation, rather than dioxygen activation. Clearly in many catalytic processes, activation of both dioxygen and substrate by the catalyst must be considered.

Selectivity is a major aim in designing artificial catalysts and consequently a knowledge of the oxidation routes used by enzymes should provide an important stimulus for catalyst design. On the other hand enzymes carrying out activation of molecular oxygen are among the least understood of enzymatic processes, primarily because of a lack of adequate chemical analogies. Research in homogeneous catalysis therefore can provide essential information to the enzymologist if oriented with some knowledge of the enzyme processes. We have attempted to survey some of the current problems in the three areas of catalysis and to find a common basis for an understanding of oxygen activation.

DIFFERENT OXYGEN SPECIES AND THEIR REACTIVITY

One major area of mutual interest to workers in various types of oxygen catalysis lies with the description of intermediates involving oxygen. Oxygen species corresponding to different states of reduction of oxygen are listed below (Table II). Many of these species have been observed directly in solution by use of techniques such as pulse radiolysis and ESR spectroscopy.

TABLE II

Oxygen Species

E^0	Species
- 0.59 V	Singlet excited species $^1\Delta, ^1\Sigma$ atomic oxygen
0.00 V	dioxygen
	$O_2^- \rightleftharpoons HO_2$
+ 1.68 V	
	$O_2^{2-} \rightleftharpoons HO_2^- \rightleftharpoons H_2O_2$
+ 0.8 V	
	$O^- \rightleftharpoons HO$
+ 1.23 V	
	$O^{2-} \rightleftharpoons OH^- \rightleftharpoons H_2O$

Singlet oxygen species are produced both by mild photoexcitation of dioxygen and in many chemical reactions in which dioxygen is a product. In the latter case part of the reaction energy may be 'diverted' into excitation energy of singlet oxygen. (It has also been suggested that certain $M\text{-}O_2$ complexes can be considered as resembling complexes of excited oxygen species). Singlet oxygen species carry out very stereoselective oxygenation reactions with allylic and diene hydrocarbons. Such species may participate in some biological, inorganic and surface reactions. The oxygen atom probably participates in some surface reactions. In most enzymatic and a few chemical hydroxylation reactions, oxygen is inserted into organic molecules as if the activated oxygen species was transferring oxygen atoms (OXENE hypothesis).

A series of dioxygen complexes of transition metals has been described. The X-ray structures of phosphine-transition metal complexes indicate that oxygen binds in a symmetrical, 'side-on' configuration. Modification of the other ligands so that more electron density is donated to the oxygen induces only an elongation of the O-O bond and does not provide a reactive oxygen species. A related problem of biological interest is the configuration adopted by the iron-oxygen adducts of proteins such as haemoglobin and cytochrome *P*-450. It is particularly relevant to ask how much electron density is transferred to the coordinated oxygen and whether a 'side-on' or an 'end on' configuration is adopted. A correlation of the properties of chemisorbed oxygen on metal and metal-oxide surfaces with those of the corresponding metal ion-oxygen complexes would seem worthwhile. Complexes of oxygen with metals may undergo oxygen-transfer reactions with either other ligands or with external substrates. An example is the reaction:

$$Pt(PPh_3)_n + O_2 \longrightarrow (Ph_3P)_2Pt(O_2) \xrightarrow{2PPh_3} 2Ph_3PO$$

The scope of such reactions has yet to be determined, but there is clearly a possibility of control over selectivity and stereospecificity.

Complete electron transfer to molecular oxygen liberates O_2^- which has been characterized in solution and on catalyst surfaces using ESR spectroscopy. ^{17}O substitution was used in studies to provide a more precise description of the radical in various environments. Extension of such work should be particularly fruitful in investigating the binding sites for O_2^- on catalyst surfaces. ESR measurements indicate that O_2^- is relatively unreactive in aqueous solutions with potentially oxidizable molecules such as alcohols and benzene. The stability of O_2^- towards disproportionation increases very considerably as the pH is raised.

The redox potential of the O_2^-/O_2 couple is also pH dependent, since the pK_a for HO_2 is 4.8. The recent determination of these properties by means of pulse radiolysis and fast reaction techniques has underlined the possibilities for work in this area.

Recently an ESR investigation of Co-Schiff base -O_2 and Co-heme-O_2 complexes has indicated that the oxygens carry 0.8 unpaired electrons. Thus a wide range of such complexes approximate to Co^{3+}-O_2^-. One mononuclear Co complex has been shown by X-ray crystallography to adopt the asymmetric 'end on' configuration. Such complexes appear to be more reactive in oxygen-transfer reactions than the symmetrical Ir complexes; e.g. such complexes carry out epoxidation of some activated olefins via a free radical mechanism. Factors controlling the reactivity of oxygen activated in this way (solvent, ligands, etc.) are at present relatively little understood. Efforts are currently being made to design ligands which by steric restrictions force the complexed oxygen to change configuration. The binuclear ions $[L_5Co\text{-}O_2\text{-}CoL_5]$ have also been shown by physical measurements to approximate to complexes of O_2^{-}-. The kinetics for formation of binuclear-metal-oxygen complexes and characterisation of products are insufficiently understood. In addition, such species may also prove useful as catalytic or stoichiometric oxidizing reagents.

The unusual selectivity of silver catalysts for epoxidation of ethylene probably results from end-on binding of dioxygen to the silver surface. Isotope-labelling techniques indicate that the terminal oxygen is inserted into the olefin. The 'end-on' configuration of dioxygen is thought to be induced by the binding of added chloride to adjacent sites so as to prevent 'side-on' binding. This process clearly has close analogies in the chemistry of solution complexes and emphasizes the value of a common approach.

In the next stage of reduction hydrogen peroxide is a relatively stable molecule. However, the deprotonated form of hydrogen peroxide, O_2^{2-}, may be formed in aprotic metal-containing sites such as hydrophobic enzyme centres and on catalytic surfaces. In the former case O_2^{2-} may be important as a reactive intermediate in enzymatic hydroxylation. Both mono and binuclear complexes are known with the following structures:

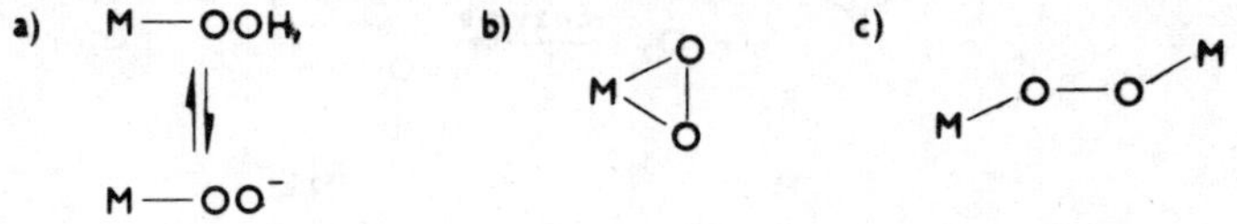

a) M—OOH ⇌ M—OO⁻ b) M<(O—O) c) M—O—O—M

Peroxo complexes of type (b) are differentiated from complexes of dioxygen by the peroxidative activity of the former, and the ready oxygenation-deoxygenation equilibrium of the latter. Peroxo complexes associate with formally high valence states and oxygen complexes with low valence states. Complexes of type (b) have been used in homogeneous catalysis, e.g. epoxidation results from Mo complexes of the type:

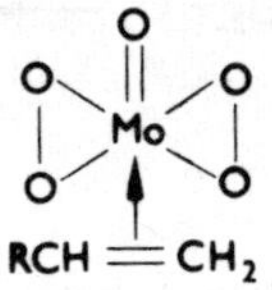

O, O, O, O, O, Mo, RCH ═ CH₂

Binuclear peroxide complexes of Co are also reactive in some instances in oxygen transfer to ligands. Such reactions appear to be facilitated by an incomplete cobalt coordination sphere. The oxygen complexes of Cu seem to be particularly worthy of further study since different geometries are preferred by the oxidized and reduced states (Cu^{I} tetrahedral, Cu^{II} planar-octahedral). The addition of oxygen to a Cu(I) ion in close proximity within a protein is an important mechanism for the enzymatic activation of oxygen, e.g. in tyrosinase or in dopamine hydroxylase. The possibility exists within a protein for the expansion of the coordination sphere of Cu upon binding oxygen by taking up further ligands from the protein. It may be significant in this respect that oxygen binding to dopamine hydroxylase and the oxygenase activity are enhanced by the interaction of fumarate with the protein.

Hydroxyl radicals and possibly corresponding metal-ion complexes are formed in the 1 e^- reduction of hydrogen peroxide. These species have been characterized by their highly electrophilic character in the oxidation of organic compounds. The equivalent anionic species probably occurs on metal oxide catalysts.

We have attempted to draw attention to the possibilities of relating active oxygen species in the different forms of catalysis. The subsequent discussion of the reactions of reactive oxygen species has been subdivided into two parts, (a) enzymatic and homogeneous reactions and (b) heterogeneous reactions, since the latter poses many of the general problems of surface chemistry.

SOME RELATED ENZYMATIC AND HOMOGENEOUS REACTIONS

Oxidative Splitting of Double Bonds

$$R_1R_2C{=}C< \; + \; O_2 \xrightarrow{\text{Enzyme}} R_1C(=O)- \; + \; R_2C(=O)-$$

This reaction can be carried out by enzymes that use dioxygen and which therefore are called dioxygenases. In the case of tryptophan pyrrolase the active intermediate is a ferroheme-O_2 complex. Absorption spectra indicate that the oxy-complex is similar in structure to oxyhaemoglobin or to compound III of peroxidase.

A similar homogeneous reaction has been reported in Co-catalysed oxidation of a small peptide:

$$\text{Glycyltryptophan} + CoCl_2 + O_2 \longrightarrow C_6H_4(C(=O)R)(NH{-}CHO)$$

where

$$R = CH_2CH(NHGly){-}COOH$$

Although no detailed $^{18}O_2$ studies have been reported it is possible that both oxygen atoms come from molecular dioxygen.

A rather different type of oxygenation mechanism occurs in catecholase enzymes which usually contain several Fe(III) ion sites. Binding of the catechol causes changes in the magnetic properties of the enzymes whereas addition of oxygen results in rupture of the catechol ring. The changes at the metal in such catalysis are obscure and further study of this type of enzyme should prove interesting. Understanding of the enzyme might be aided by further study of the catechol-Fe^{3+}-O_2 and related systems.

Homogeneous catalytic oxidation in which metal-coordinated dioxygen reacts catalytically using both oxygen atoms have been described.

$$L_nM(O_2) + 2L \longrightarrow L_nM + 2OL$$

$$L_nM + O_2 \longrightarrow L_nM\langle{}^{O}_{O}$$

$$L = PR_3,\ RNC$$

$$M = Pt,\ Pd,\ Ni,\ Rh^{I},\ Ru^{II},\ Co^{III},\ etc.$$

Intermediates such as

$$L_nM(O_2)(L)_2 \longrightarrow L_nM(OL)_2 \quad \text{or} \quad L_nM(O{-}L)(O)(L) \longrightarrow L_nM(OL)_2$$

have been proposed.

There is another class of enzymes which introduce molecular oxygen into highly unsaturated systems like polyunsaturated fatty acids, possibly without metal activation. These substrates are reactive enough to allow either addition of dioxygen to the double bonds or insertion into CH bonds. Reactions of this type occur in lipoxygenases and prostaglandin synthesase.

The ability of Co(II) (and also Cu(II), Mo(II), Fe(II)) to initiate radical chains may involve complexing of dioxygen to produce a radical dioxygen capable of hydrogen atom abstraction from the substrate:

$$Co(II) + O_2 \rightleftharpoons Co(III)\,O_2^{-}\cdot$$

$$Co(III)\,O_2^{-}\cdot + RH \longrightarrow Co(III)\,OOH + R\cdot$$

A radical chain reaction then takes place as follows:

$$R\cdot + O_2 \longrightarrow RO_2\cdot$$

$$RO_2\cdot + RH \longrightarrow ROOH + R\cdot$$

$$ROOH \longrightarrow RO\cdot + OH\cdot \ \text{etc.}$$

Oxidation of unsaturated fatty acids, aromatic compounds and saturated hydrocarbons may be thus achieved.

Hydroxylations or Mono-Oxygenation

One of the most important features of hydroxylation in biological systems is to provide a means of introducing a functional group into a hydrocarbon molecule. The use of ^{18}O has indicated that the oxygen atom introduced into the substrate stems from the dioxygen molecule whereas the second atom is reduced to water. The monooxygenases also require an external electron donor according to the equation:

$$RH + {}^{18}O_2 + H^+ + 2e \longrightarrow R^{18}OH + H_2O^{18}$$

Many enzymes of this type contain Fe or Cu at the catalytic site. However there is also a substantial class of hydroxylanes in which oxygen is activated by flavin, possibly as a flavin hydroperoxide. These hydroxylanes generally have more reactive phenolic substrates.

Attempts have been made in homogeneous catalysis to simulate these reactions, partly to get information about the principles of oxygen activation and partly to find a method for the chemical hydroxylation of organic compounds. The chemical features of the enzymatic aromatic hydroxylation include electrophilic substitution patterns and the ready migrations of substituents from the position of substitution. Both features have been shown to be consistent with the intermediate formation of aromatic epoxides. In this respect, the enzymatic monooxygenation proceeds in a fashion similar t peracid oxidations so that an 'oxenoid' mechanism involving a Fe(III)peroxo complex has been proposed.

The difficulties in simulating these enzymatic processes in homogeneous solution derive partly from the fact that the reducing agent will compete very effectively for the active oxygen reducing it to OH-radicals or water. It may be worthwhile to seek catalytic systems in which electron donation must exclusively occur through the metal ion and be restricted to a two-electron transfer process. In an enzyme this restriction is supplied by th selective steric requirements imposed upon the substrate and reducing agent

The currently known homogeneous model systems in general involve predominantly radical processes. Further detailed study seems necessary of known i organic hydroxylation systems such as that based on stannous pyrophosphate, in which the metal can transfer two electrons to oxygen. Cu(I)-catalysed hydroxylations of phenols in morpholine produce specific ortho-attack. It may be worth looking for other specific systems of this type.

Oxygen-Activated Dehydrogenation of Aliphatic Compounds

Usually the enzymatic formation of a double bond in saturated hydrocarbon is achieved by dehydrogenation. This, however, will become energetically unfavorable when CH-activating groups are lacking. In these cases, e.g. long-chain fatty acids, the dehydroxygenation occurs by a two-electron activation of the dioxygen molecule very similar to that in monooxygenase reactions. In contrast to the monooxygenases the desaturases will use the activated oxygen species only as hydrogen acceptor according to the equation:

$$\begin{matrix} \diagdown CH_2 \\ | \\ \diagup CH_2 \end{matrix} + [O] \longrightarrow \begin{matrix} \diagdown CH \\ \| \\ \diagup CH \end{matrix} + H_2O$$

The alternative mechanism via hydroxylation and subsequent dehydration has been ruled out. No attempts to simulate this interesting reaction in a homo geneous system have been successful mainly because of the same difficulties as in the models for hydroxylation. However, this type of reaction takes place readily on some transition-metal oxides at high temperatures, possibly by a similar hydrogen abstraction.

SUGGESTIONS FOR FUTURE RESEARCH

Homogeneous and Enzyme Catalysis
(some applicable to heterogeneous systems)

(1) Methods for producing and studying unstable metal ion-O_2 complexes.

(a) Structural restrictions and other factors which convert 'side-on' to 'end-on' dioxygen complexes.

(b) Methods of activating dioxygen complexes. Stable metal-dioxygen complexes in enzymes are activated by further reduction in an aprotic region of the enzyme, probably with the substrate bound close to the dioxygen. Models aiming at any of these features might provide the basis of an active homogeneous catalyst and give insight into the enzyme mechanism. Photolytic activation might also be worth studying.

(c) The study of a wider range of metal ion-oxygen reactions, particularly in non-aqueous media, both for kinetic and mechanistic data and as a source of reactive intermediates for the oxygenation of organic substrates. The use of pulse radiolysis may be useful in increasing the versatility of such a study. For such studies it will be essential to have several criteria for resolving substrate attack by different species. The use of ^{18}O-labelled dioxygen to trace the source of oxygen incorporated into substrates is very important.

(2) A thermodynamic study of the different states of reduction of dioxygen.

(3) Investigation into the oxygen-transfer step in hydrocarbon oxygenation, particularly with reference to changes at the hydrocarbon such as the formation of metal-carbon species. Activation of oxygen cannot be considered in isolation from activation of the co-reactants.

(4) We have generally discussed methods of activating oxygen. It is clearly essential to examine approaches towards making such homogeneous reactions truly catalytic by providing an appropriate supply of reducing equivalents. Research into electrolytic reduction in this field would seem particularly appropriate.

Points 1(a), 2 *and* 3 *are clearly in many respects applicable to heterogeneous catalysis*. It should be pointed out that photoactivation of heterogeneous oxygenation has already proved successful.

Heterogeneous Catalysis

Certain recommendations specifically concerning the study of heterogeneous oxygen catalysis are made below. In general we feel that it is important to encourage a physically and conceptually oriented approach to the subject. The following general points may help in consideration of oxygen catalysis but are clearly applicable to surface catalysis as a whole.

(1) The availability of standardized oxide catalysts which have been characterized physically and in chemical reactivity towards test reactions.

(2) To relate catalysis to a knowledge of sites which can be characterized by geometry, energetics (calorimetry etc.), and spectroscopy (particula ly ESR and IR of surface-bound groups).

(3) Many oxidation states of oxygen and dioxygen can exist on catalyst surfaces and therefore may participate in oxidation reactions. It seems important to find means to characterize such species.

(4) To compare reactivities of surface oxygen species with simple molecules and then to proceed to increasingly complex processes, e.g. isotopic exchange of molecular oxygen with active surface oxygen has been shown to adopt two or more different mechanisms on various oxide surfaces.

(5) A correlation of catalytic properties with the chemical properties of ions in a surface environment, e.g. changes in valency states particularly to unusual valence states which may be stabilized in the surface. It seems important to obtain an understanding of the distribution of 'multicentre sites' which may have unusual reactivity patterns. The use of the knowledge of species formed in homogeneous metal-ion catalysis should be va valuable in assessing intermediate surface complexes in heterogeneous processes.

(6) The effect of oxygen mobilization through dynamic cooperative phenomena such as local shear structures which may be induced by added ions or mixed oxides.

(7) It is clearly important to relate studies of activated oxygen species to active organic species present in the surface.

GENERAL SUMMARY

The experience gained in these discussions has emphasized the value of interrelating efforts in the three fields of catalysis. The following *sugges tions* are highlighted for future research:

(1) Development of means for reducing metal ions and metal-ion-oxygen complexes (e.g. electrolytic) so that the oxidation of substrates can be made catalytic with respect to the metal ion.

(2) Methods for producing and studying *unstable* metal-oxygen complexes and surface species, including an extension of the range of metal ions exam ined and the study of means of activating stable dioxygen complexes, e.g. reduction by photolysis.

(3) A study of factors controlling 'side-on' and 'end-on' dioxygen complexe in metal-ion complexes, metalloenzymes and on surfaces.

(4) Investigation of possible intermediate organometal, and organometal-oxy gen species which may occur in hydrocarbon oxygenation, and control of the products obtained.

(5) Encouragement of further attempts to assess heterogeneous processes using the knowledge of the reactivity of metal ions and metal-oxygen species gained from solution chemistry.

HYDROCARBON ACTIVATION

CHAIRMAN: G. W. Parshall

E.I. du Pont de Nemours & Company, Central Research Department, Experimental Station, Wilmington, Delaware 19898, USA

RECORDER: M. L. H. Green

Inorganic Chemistry Laboratory, University of Oxford, South Parks Road, Oxford OX1 3QR, England

MEMBERS: F. Calderazzo D. A. Dowden F. G. Gault J. Halpern G. A. Hamilton C. Kemball S. J. Teichner H. C. Volger

In view of the dwindling supply and increasing cost of natural gas and petroleum, the prime sources of energy for our society and feedstocks for our chemical industries, it is vital that we use them with utmost efficiency. Major contributions to more efficient use can be made by improving the selectivity with which industry carries out transformations of saturated hydrocarbons. Even minor improvements in the effectiveness of catalytic cracking and reforming can mean immense quantitative savings in crude hydrocarbon consumption. Improvements which lead to higher-quality motor fuel should also reduce the total amount of fuel consumed with attendant reduction of air pollution.

In the following pages, the status of research with respect to hydrocarbon activation is summarized and areas for future research are considered in terms of their practical application and theoretical significance. Given the diverse areas of strength amongst heterogeneous, homogeneous, and metalloenzyme catalysis, the opportunities for progress through multidisciplinary research seem attractive.

CURRENT STATUS IN RESEARCH

Heterogeneous Catalysis

The activation of saturated hydrocarbons is carried out on an immense scale in the chemical industries and in petroleum refineries but understanding of the catalytic processes involved has lagged far behind commercial application. The interactions of C-H and C-C bonds with surfaces are at the heart of such widely applied reactions as isomerization, cracking, reforming, and steam-reforming of alkanes. In recent years considerable progress has been made in learning about such interactions on some metal surfaces and also on a number of oxides, but there is need for studies on a wider range of surfaces and a development of the fundamental concepts involved.

Current research related to the theme of the activation of saturated hydrocarbons by surfaces takes many forms. Adsorption studies continue to be made on a range of solids and are amplified by the use of infrared spectroscopy. Recent work with well-defined metal surfaces using ultra high vacuum and a battery of physical techniques such as LEED, Auger spectroscopy and X-ray photoelectron spectroscopy provides important basic information. Exchange reactions of alkanes with deuterium or sometimes deuterium oxide supply valuable data on the rates of formation of reversibly formed intermediates on a range of surfaces and about the types of reactions which these intermediates can undergo. Kinetic studies, sometimes assisted by isotopic labelling techniques, are in progress on many categories of reactions import to industry. These include cracking and skeletal isomerization, dehydroge ation and oxidative dehydrogenation, hydrogenolysis, dehydrocyclization an other reforming processes, and chlorination. Acidic oxides, particularly zeolites, non-acidic oxides, metals, and other types of catalysts such as sulfides feature in many of these studies.

Metalloenzyme Catalysis

The activation of saturated hydrocarbons by biological systems occurs fr quently in normal metabolism and is potentially important in the productio of food from petroleum. It is obvious that the organisms which grow with alkanes as the sole source of carbon have the ability to convert alkanes t functionalized derivatives. The reaction of unactivated carbon-hydrogen bonds also occurs in many other metabolic pathways.

The most common reaction of alkanes in biological systems is the conversion of the alkane to an alcohol with molecular oxygen as the oxidant. Th overall stoichiometry of this reaction is: $RH + O_2 + AH_2 \rightarrow ROH + A + H_2O$, which AH_2 is one of a variety of biological reducing agents. Another freq ently occurring reaction, however, is the direct dehydrogenation of an alk to an alkene. Again oxygen is the oxidant and the overall stoichiometry i

$$\text{Alkane} + O_2 + AH_2 \longrightarrow \text{alkene} + 2H_2O + A$$

It is significant that hydroperoxides are *not* intermediates in either of these reactions, and in fact there is no authenticated example of enzyme-catalysed hydroperoxide formation from an alkane.

Both these reactions apparently proceed by reaction of the O_2 and reducir agent (AH_2) with the enzyme to give a very reactive oxygen species. This species then reacts with an alkane molecule bound only by noncovalent bonds to the enzyme. The alternate pathway of alkane activation followed by reac tion with O_2 does not appear to occur in these reactions. The reasons for the first rather than the second mechanism are probably thermodynamic. The high energy inherent in the O_2 molecule can be harnessed under physiological conditions to react with an unreactive alkane, but the energy required to convert an alkane to a highly reactive intermediate would not be available to the enzyme. Some enzymatic reactions which utilize the coenzyme B_{12} (which contains Co) probably proceed by activation of an alkyl group by the enzyme, but these reactions result in rearrangement rather than substitution of the hydrocarbon.

Homogeneous Catalysis

The reported examples of homogeneous activation of saturated hydrocarbons by metal complexes are few. A system based on a $PtCl_2$-$PtCl_4$ couple in aqueous acid is reported to activate alkane C-H bonds toward exchange with D_2O and to chlorination. Strained cyclic hydrocarbon C-C bonds such as those o cyclopropane and cubane are broken by complexes of Rh, Pt, Ag, and other metals. Chromyl chloride, although best known as a selective oxidizing agent for activated hydrocarbons, oxidizes paraffins. Practical application of these reactions does not seem to have been reported.

CONCLUSIONS

General

The discussions held at this conference have clearly indicated the need for interdisciplinary communication if progress in the catalytic activation of saturated hydrocarbons is to occur as fast as possible. To date, there has been little discussion across the interfaces between homogeneous, heterogeneous, and metalloenzyme research. We strongly recommend a systematic attempt to promote communication in the interfacial areas, particularly the enzymatic-homogeneous and homogeneous-heterogeneous interfaces.

The character of work required in each of the three catalytic disciplines is different. Thus, while chemists in homogeneous and heterogeneous catalysis can evolve new processes, in the metalloenzyme area one can only discover and study pre-existing processes to obtain an understanding. The knowledge gained in this way can be very useful in design of catalytic systems that operate under mild conditions. Conversely, the principles and models developed in homogeneous and heterogeneous catalysis can be applied to the understanding of biological systems. Thus we recommend closer contact between researchers in the metalloenzyme field and those in the more traditional catalysis areas and that they make an effort to convey their findings and conclusions to their colleagues in other fields.

The great practical achievements in heterogeneous catalysis present a tremendous stimulation to our quest for a better understanding of the processes involved. In heterogeneous catalysis, the main problems are not so much concerned with the initial activation of the C-H bond as with the control of the subsequent processes. These may often occur too rapidly, leading either to degradation of the alkane to small molecules such as methane, or alternatively to the formation of polynuclear aromatics and coke. On the other hand, a more fundamental knowledge of the process of C-H bond rupture at metal surfaces should contribute to progress in the homogeneous activation of hydrocarbons.

Homogeneous catalysis of unstrained, saturated hydrocarbon reactions is scarcely known. Achievements in this area would have great potential for improved selectivity in converting alkanes to functional derivatives such as alcohols, ketones, acids, nitriles, amines, and olefin derivatives. We recommend major efforts to develop this area, drawing heavily on knowledge obtained from heterogeneous and metalloenzyme catalysis. In keeping with this recommendation, the suggested areas for research tabulated below emphasize work in homogeneous catalysis and its interfaces with the other two approaches.

SPECIFIC AREAS FOR RESEARCH

I. Activation of the C-H Bond

A. *Heterogeneous Catalysis*

Studies of exchange reactions of alkanes with deuterium or of self-exchange (e.g. CH_4 with CD_4) have provided information about the rate of reversible dissociation of hydrocarbons on a number of surfaces. Most of the results refer to metals, a few alloys, and a rather limited range of oxides. The following two areas appear to merit attention.

1. Alkane Exchange on a Range of Surfaces

Any class of solid of actual or potential interest from the industrial o the academic point of view should be examined for exchange reactions of al anes. Immediate possibilities include zeolites, sulfides and a number of alloys or mixed oxides.

2. Attempts to Obtain a More Detailed Picture of the Process of Dissociative Adsorption on Metals

The process of dissociation of an alkane on a metal is normally written

$$RH + 2M \longrightarrow \underset{|}{R}\!\!-\!M + \underset{|}{H}\!\!-\!M$$

where M refers to a metal atom or an adsorption site. An alternative formulation suggested by the behaviour in homogeneous catalysis is

$$RH + M \longrightarrow R\diagdown M \diagup H$$

possibly followed by migration. Deciding between these possibilities will not be easy, but the distinction could be important because it is highly relevant to the interface between heterogeneous and homogeneous catalysis. It is obviously related to the problem of a better description of the catalytic sites on metals (*vide infra*).

B. *Homogeneous Activation of Alkanes*

1. Direct Synthetic Approach

We recommend that the *best approach to homogeneous* C-H *activation* is broa imaginative synthetic study based on induction from related chemistry.

(a) *Homolytic activation.* The reversible oxidative-addition under mild con ditions of C-H bonds in saturated hydrocarbon groups which, as ligand substitutents, are held near to transition-metal centers is well-known (the proximity effect). The general reaction may be written as:

$$M-L-(C)_n-H \longrightarrow H-M(-L)-(C)_n \quad \text{(ring: } M\text{–}L\text{–}(C)_n\text{–}M\text{)}$$

Clearly the electronic requirements for C-H activation are available in discrete transition-metal centers. Induction from known oxidative-addition chemistry suggests that the best catalysts will contain very electron-rich metal centers which are coordinatively unsaturated. Since existing compound do not apparently achieve activation of free alkanes it seems that new or un familiar environments need to be explored.

One limitation on catalyst design is the 'proximity effect' which limits ligand choice since C-H groups on the ligands must not compete preferentiall with free alkane C-H. For example, the complex $(\pi\text{-}C_5H_5)_2.NbH_3$ readily catalyses H/D exchange in benezene, but, in alkane solutions, a C-H bond of the

π-C_5H_5) ring is preferentially attacked to give an inert binuclear product.

A different approach is suggested by analogy with enzyme sites in which ighly energetic metal centers are associated with unconventional configur- .tions. For example, metal porphyrin or corrin systems can be reduced to ive very electron-rich centers and the metal is often accessible along he out-of-plane axis. The known chemistry of the 17-electron system $Co(CN)_5^{\cdot})^{3-}$ suggests an approach involving activation of alkanes *via* a homo- ytic process. An example, in principle, is: $(Co(CN)_5^{\cdot})^{3-}$ + RH → $Co(CN)_5H$ R·. Such a radical process is expected to have little activation energy n excess of endothermicity. In this hypothetical example, since the Co-H ond is weaker than the C-H bond, the equilibrium position has the backward lirection. However, in other systems, a strong M-H bond should promote the eaction. In this context, attention is drawn to the role of Co in vitamin $_{12}$ reactions with alkyl-containing substrates.

b) *Heterolytic* C-H *cleavage*. The following reactions may be envisaged:

$$\text{Strong electrophile } M^+ + C\text{—}H \begin{cases} M\text{—}C + (H^+) \\ M\text{—}H + (CH_3^+) \end{cases}$$

$$\text{Strong nucleophile } M^- + C\text{—}H \begin{cases} M\text{—}H + (R^-) \\ M\text{—}R + (H^-) \end{cases}$$

Since alkanes have a filled shell, the reactions with strong electrophiles seem more likely. Further, it seems probable that a bifunctional catalyst would be required and that analogy with the known activation of C-H bonds by metal-oxide surfaces might be useful in this context.

(c) *Other approaches*. Syntheses of very highly reactive catalysts for alkane activation may require less common synthetic approaches — e.g. electrochemical or photochemical methods. Thus the thermally inert system, $(\pi\text{-}C_5H_5)_2WH_2$ in benzene -d_6 gives the phenyl derivative, $(\pi\text{-}C_5H_5)_2WD(C_6D_5)$ in 70-80% yield at 5 °C on irradiation with visible light. Metal-vapour synthesis could also be a valuable synthetic approach. An example of this method is:

$$2C_6H_6\ (\text{vapor}) + Mo\ (\text{vapor}) \xrightarrow{-195°} (C_6H_6)_2Mo$$

2. Model Systems

(a) A study of the synthesis and properties of *cis*-alkylhydride and *cis*-di- alkyl metal derivatives would be desirable, especitally experiments des- igned to reveal the factors affecting their stability. These compounds may serve as models for chemisorbed alkanes on surfaces.

(b) There are relatively few thermodynamic data concerning transition-metal bonds to carbon or hydrogen ligands, which is at least in part due to the difficulty of obtaining such data. Nonetheless, we feel that greater understanding of both heterogeneous and homogeneous alkane catalysts requires *more thermodynamic data* on pertinent systems.

II. Activation of Saturated Carbon-Carbon Bonds

The important objectives in the activation of saturated hydrocarbons include catalysis of reactions involving the dissociation or rearrangements c carbon-carbon bonds.

Such catalytic processes are well known; for example, acid-catalysed isomerizations of alkanes with both homogeneous and heterogeneous acid catalys as well as heterogeneous catalytic hydrogenolysis reactions. Investigation of such hydrogenolysis and isomerization reactions are beginning to reveal the complexities of the processes associated with the rupture of C-C bonds and the skeletal rearrangements of hydrocarbons in the heterogeneous catalysis. For example, in acid-catalysed reactions, it is likely that the rearrangements proceed through intermediate carbonium ions.

Most of the reactions cited above are probably initiated by C-H, rather than C-C bond dissociation (at least the *initial* role of the catalyst is th activation of a C-H bond). Nevertheless, the exploration of possible catalytic approaches to direct activation of saturated C-C bonds seems importan In this connection, the recent discovery of the catalytic disproportionatio of olefins revealed a highly unexpected mode of dissociation of the latter molecules, i.e. dissociation of the very strong C=C bond rather than one of the C-H bonds.

A. *Approaches Related to* C-H *Activation*

The most promising approaches to the direct activation of saturated C-C bonds are similar to those already elaborated for C-H bonds. In the case o homogeneous systems these involve the use of metal ions and complexes to effect oxidative addition or heterolytic cleavage of the C-C bond, e.g.:

$$M + R{-}R' \longrightarrow M\begin{matrix} \diagup R \\ \diagdown R' \end{matrix}$$

$$M^+ + R{-}R' \longrightarrow M{-}R' + R^+$$

It is recognized that (for kinetic rather than thermodynamic reasons) such processes are probably more difficult than in the case of H-H or C-H bonds and require more reactive catalytic complexes. However, the approaches that might be fruitful for the development of such more reactive catalysts seem basically similar to those suggested for C-H activation. It is recommended that the lines of investigation proposed for the latter be pursued also in the specific context of the activation of C-C bonds.

B. *Rearrangement of Strained Hydrocarbons*

The catalytic activation and rupture of saturated C-C bonds through oxidative addition and heterolytic dissociation are already well known for *strain* hydrocarbons, notably those containing cyclopropane and cyclobutane rings. In recent years, numerous instances of the catalytic rearrangement of such hydrocarbons by transition-metal compounds have been described; for example, the Rh(I)-catalysed rearrangements of quadri-cyclane to norbornadiene and of cubane to *sym*-tricyclo octadiene, as well as the Ag(I)-catalysed rearrangements of cubane.

There are noteworthy parallels between the modes of reaction of C-C bonds with these two different ions and the corresponding reactions of the ions ith H_2. Thus Rh^I and other d^8 complexes typically dissociate H_2 through xidative-addition reactions whereas Ag^I has been shown to dissociate H_2 hrough heterolysis, e.g.

$$Rh^I(PPh_3)_3Cl + H_2 \longrightarrow \begin{matrix} H \\ H \end{matrix}\!\!>\! Rh^{III}(PPh_3)_3Cl$$

$$Ag^I + H_2 \longrightarrow Ag^IH + H^+$$

This parallel suggests that at least some features of the activation of different types of saturated bonds are similar and affords some support for a unified approach to their investigation. Thus, it is recommended that investigations directed at the activation continue to include studies of the rearrangements of strained hydrocarbons, notably from the standpoints of (i) learning more about the activity patterns (e.g. influence of metal and ligand variation upon catalytic activity) that govern the activation of C-C bonds in such systems and (ii) systematically extending such studies to the rearrangements of less strained hydrocarbons.

III. Homogeneous and Enzymatic Oxidation Studies

A. *Homogeneous Studies Stimulated by Enzymatic Information*

Free-radical, metal-catalysed and heterogeneous oxidations of paraffins by O_2 as well as chlorinations by Cl_2 are usually unselective. Since metal-containing enzymes usually functionalize alkanes by selective oxidation with O_2, it seems profitable to attempt to mimic both the selectivity and mechanism of the enzymatic process in homogeneous systems. It is recommended that attempts be made to modify the selectivity of these reactions by changing the reactivity of the hydrogen-abstracting species. This might be done by altering the steric requirements of the reagent or by complexing it to various metal centers. In these cases medium to high oxidation states for the central metal atom may be desirable.

In recent years, several synthetic metal complexes of dioxygen have been reported in the literature, mainly with Co, Pt, and Ir. Both the 1:1 (metal: O_2) and the 2:1 types of complexes are known, whereas the only well recognized metal-containing enzymatic type is the 1:1. It seems that a thorough investigation of the reactivity of the synthetic complexes, especially of the 1:1 type, may lead to a selective autoxidation of C-H bonds. It should be recognized, however, that only very unstable O_2 complexes are expected to be reactive enough to attack alkanes. Fe-dioxygen complexes which until now have been rather elusive to preparative chemists appear particularly interesting because the known enzymes which catalyse the alkane to alcohol reaction contain Fe. The enzymatic reaction apparently proceeds by the direct insertion of an oxygen atom into the carbon-hydrogen bond by an oxenoid mechanism. Thus, a search for homogeneous catalytic systems which might be expected to perform a similar reaction is indicated.

B. *Homogeneous Studies Stimulated by Results from Heterogeneous Catalysis*

The adsorption and oxidation of paraffins on metal-oxide surfaces suggest that single-phase studies with predominantly covalent metal oxides (Re_2O_7,

MoO_3, WO_3, $CrO(C_5H_5)_4$) may lead to a better understanding of the interactic of C-H bonds with M-O bonds. Possibly the high polarity of the M-O bond wi induce polarization of the C-H bond, thus promoting both the approach of th reactants and the insertion of oxygen into the C-H bond. A similar type of reasoning should also apply to the previously mentioned attempts to chlorin ate alkanes by a selective, non-free-radical process.

C. *Soluble Enzyme Studies*

Further clarification of the enzymatic oxidations of alkanes is needed. Most enzymes which have been studied are membrane-bound, and thus insoluble in water. This has stymied progress in determining the details of the reac tion mechanism. Major effort should be devoted to finding and characterizi soluble enzymes which perform the same function. Microorganisms and fungi, rather than mammals, are likely sources for such enzymes. If such enzymes are found they should be studied by the sophisticated kinetic and physical methods available for the study of soluble enzymes, including spectral and magnetic resonance characterization of intermediates. Complete structure determination by X-ray methods may be justified.

IV. Metal Aggregates as Catalysts

A large gap exists between the approaches in homogeneous and heterogeneou catalysis. In the former, the emphasis is mainly focussed on mononuclear complexes in various oxidation states and coordination numbers. In heterogeneous catalysis attention has focussed on metals and oxide and sulfide crystals.

In heterogeneous catalysis, a variety of reactions proceed simultaneously on a given catalyst and high selectivity is very difficult to achieve. Attainment of similar reactions with homogeneous catalysts might be expected to lead to greater selectivity.

In order to increase the selectivity in heterogeneous catalysis and to per form some reactions of saturated hydrocarbons not yet observed in homogeneou catalysis, it is felt that research in an intermediate area between large crystal and mononuclear complexes on better defined catalytic systems is ver desirable.

(A) Starting from the heterogeneous viewpoint, the interrelation between the selectivity and the size of the crystallites is a prime candidate fo study, especially in the range below about 15 Å. A possible way to produce much smaller crystallites would be to use supports such as synthetic or natural fibers or resins.

In homogeneous catalysis, we suggest a systematic investigation of the catalytic properties, activity and selectivity of well-defined clusters. Another desirable approach is the investigation of well-defined homogeneous complexes bound to suitable resins and fibers. In such heterogeneous systems, the arrangement around the metal atom and the bond distances could be established.

(B) A second valuable approach to fill the gap between both types of catalysis would be the use of heteronuclear crystallites and clusters. In heterogeneous catalysis, alloying a metal changes its catalytic properties. In the homogeneous field it appears that novel types of reactions can be expected from use of catalytic systems involving two or more different metals. It will be of interest to study the catalytic properties of bimetallic mixed clusters in homogeneous catalysis, and conversely to investigate the structure and catalytic properties of very small crystals consisting of two or more metals, and, more generally, of two or more metallic compounds.

V. Controlling the Activity of Heterogeneous Catalysts

Modification of the activity and the selectivity of heterogeneous catalysts by the addition of specific poisons and inhibitors is much practised in academic research and in industry, although not always affording the desired degree of control. Customarily, metals have been selectively poisoned by substances (e.g. organic sulfides) added to the reactants, or by inactive metals and non-metals added to the catalyst during preparation, e.g. lead in Lindlar's catalyst. In multicomponent catalysts, the undesirable reactions effected by some components, such as those due to the acidic functions of a support, have been inhibited by the inclusion of a specific poison, in this case a base.

The temperatures required to activate alkanes are often so high that unwanted side reactions occur and it is desirable to minimize such parasitic processes. These problems are always encountered with the active transition metals and with the strong acids.

A. *Alloyed Metals*

The very active transition metals tend to produce polymeric residues, 'coke', carbon, etc., by stripping hydrogen atoms from the alkane skeleton. Improved selectivity and stability may be achieved by alloying such active metal catalysts. Nevertheless, the systematic investigation of the effects of alloying and intermetallic compound formation, particularly with the object of controlling side reactions, has been neglected in recent years. Solid-state physics and metallurgy now provide adequate background information upon which exploratory and mechanistic research can be soundly based. Current industrial trends clearly show that alloy catalysts have become important factors in petroleum refining. It is strongly recommended that research in these topics be undertaken.

B. *Ligand Inhibitors and Activators*

The remarkable modification of the activity and selectivity of homogeneous complex catalysts induced by changes in the ligands suggests that somewhat similar effects might result from the use of more complex inhibitors on the surfaces of solid catalysts. Programmes of research based upon the investigation of a series of known, stable electron donor or acceptor ligands as inhibitors for metal and oxide catalysts using a wide range of reaction types should be revealing both for the intrinsic value of the results and for the basic information on electronic and steric effects.

C. *Acidic Catalysts*

In a similar way the more extreme activities of carbonium ions occurring in reactions over acidic catalysts (SiO_2-Al_2O_3, exchange zeolites, etc.) might be modified by the inclusion of some uncommon anionic ligands, (e.g. PF_6^-) and by other metals and cations.

Zeolites exhibit properties, particularly in the mobility of lattice and adsorbed components, which make it reasonable to classify them in the region between heterogeneous and homogeneous catalysts. Therefore, an investigation of the effects of novel ligands and other species of suitable size on the behavior of the carbonium-ion reactions is a worthwhile exercise for its own sake and for the possible synergy between homogeneous and heterogeneous catalysis.

VI. Techniques and Applications

A. *Homogeneous Catalysis at Elevated Temperatures*

(1) Some processes use as catalysts molten salts held in the pores of more or

less inert supports. Examples of such systems include the potassium sulfate-vanadium sulfate on silica catalyst for the oxidation of sulfur dioxide and the alkali halide-copper halide oxychlorination catalysts.

If the outstanding selectivity of homogeneous catalysts evident at temperatures below about 100 °C can be exploited at higher temperatures then a greater range of selective reactions might be found (e.g. dehydrogenation) and the rates of known homogeneous reactions increased. Already patents exist for the application of homogeneous catalysts as solutions in inert high-boiling solvents sustained within the pores of materials. There seems to be good reason to explore the use of molten media of still higher melting point, e.g. $Et_4N^+SnCl_3^-$ containing dissolved complexes appropriate to the desired reactions.

(2) Alternatively new complexes will be found carrying ligands which are stable at higher temperatures and which confer thermal stability upon the whole complex. Such stable complexes would be more successful in activating saturated hydrocarbons.

B. *Special Types of Supported Catalysts*

Active species from enzyme chemistry and from heterogeneous catalysis can be extended through supports to forms which are unusual for their class and which may then induce atypical activity for the reactions of paraffins.

(1) Supported enzymes may induce new types of activity in enzymes which already exhibit some activity related to alkane activation.

(2) Similarly metals and ions can be dispersed or exchanged onto acidic, neutral and basic resins to give useful multifunctional activity. Composites of this nature might be made lyophilic or lyophobic so as to be compatible with various environments.

C. *Investigation of Active Sites*

The recurrent theme of this conference has been the ubiquity of the principles of coordination chemistry. In particular we refer to the importance of electronic configuration, electronic energy, symmetry, and coordination number. Recent investigations into chemisorption and catalysis by transition metal ions dissolved in inert solid matrices have given strong indications of the presence of such effects in some kinds of heterogeneous catalysts. There are also signs that a similar approach can be made to the properties of single atoms and ensembles of atoms arising as defects in metals or engineered into alloys. The presence of steps, edges and other defects further influences the problem.

It is therefore suggested that special attention be devoted to the activation of saturated hydrocarbons first on well-defined centers in single crystal surfaces (metals, oxides, etc.), and then to 'polynuclear' centers in such surfaces, next to stepped surfaces and finally to real catalysts. Preferably a probe reaction would be chosen which is structure-sensitive, exhibits selectivity and concerning which a sufficient knowledge of mechanism exists. The object would be to discover the detailed mechanism of the process and then with this knowledge to model other heterogeneous, homogeneous and possibly enzyme catalysts.

D. *Hydrogen-Transfer Reactions*

Some enzymes catalyse hydrogen transfer from an alkane to some other substrate, in effect, an oxidative dehydrogenation. Similar reactions have been investigated using heterogeneous catalysts and will be important in the future

It is suggested that a reaction which can be catalysed by enzymes, by homogeneous gas phase catalysts and by heterogeneous catalysts should be noted as worthy of special attention in the context of the preceeding paragraphs.

E. *Photoactivation of Alkane Reactions*

It has recently been shown that the selective catalytic oxidation of some simple alkanes at temperatures between 30 °C and 100 °C over anatase is induced only by continuous irradiation with light having energy equivalent to the energy gap between the valency and conductivity bands of the oxide. Photocatalytic oxidation of this character is novel and demonstrates the existence of a powerful but unexploited technique for the production of oxygenated hydrocarbons. It forms an area of research which should be explored.

F. *Electrocatalysis*

Catalytic problems involving the activation of saturated hydrocarbons occur in fuel-cell technology and in the production of oxidized hydrocarbons by electrocatalytic methods. Already much money is spent on the science and technology related to fuel cells but more effort should be spent on the investigation of the selective oxidation of hydrocarbons at catalytic electrodes.

There exists a coupling with homogeneous oxidation because the oxidation can be carried out or initiated in homogeneous solution by dissolved cations (e.g. Ce^{4+}), the reduced forms of which are then reoxidized at the cell electrode. A place should be found for research along these lines.

SUMMARY

The relatively underdeveloped state of research on homogeneous catalytic activation of saturated hydrocarbons marks this topic as one of maximum scientific challenge and opportunity. Research in the homogeneous area seems sure to profit from exchange of information with the more highly developed areas of heterogeneous and enzymatic research on alkane activation. Information flow in the opposite direction should also be very worthwhile and should be encouraged.

We highly recommend (roughly in order of priority):

(1) Research on exceptionally reactive transition-metal complexes as a route to C-H and C-C bond activation.

(2) Synthesis and thermodynamic characterization of alkyl-metal complexes as models for intermediates in both homogeneous and heterogeneous activation of alkanes.

(3) Study of selective alkane oxidations catalysed by metal complexes modelled after metalloenzyme and heterogeneous catalysis.

(4) Convergent research in homogeneous and heterogeneous alkane activation by metal cluster complexes and by solids containing metal aggregates.

(5) Further characterization of the sites in heterogeneous catalysts for alkane activation.

(6) Selective attachment of ligands to a heterogeneous catalyst surface should be studied as a method of moderating the activity of such catalysts for alkane activation.

REFERENCES

(for general reading on catalytic activation of alkanes)

Heterogeneous

KEMBALL, C. (1959). *Adv. Catalysis*, 11, 223.

Homogeneous

HALPERN, J. and PARSHALL, G.W. (1973). See articles by these authors in *Collected Accounts of Transition Metal Chemistry*, (American Chemical Society).

Metalloenzymes

No general review has been noted; see references in the introductory article by G.A. Hamilton in this volume.

HETEROGENIZING CATALYSTS

CHAIRMAN: J. Manassen

Department of Plastics Research, Weizmann Institute of Science, Rehovoth, Israel

RECORDER: D. D. Whitehurst

Research and Development, Mobil Oil Corporation, Paulsboro', New Jersey 08066, USA

MEMBERS: J. C. Bailar, Jr. S. Carra M. Graziani K. Mosbach R. C. Pitkethly E. K. Pye J. J. Rooney

INTRODUCTION†

In this report, a heterogeneous catalyst is defined as one which can be readily separated from the reaction mixture by simple, economical, physical means; for example, by filtration or centrifugation. This ease of separation gives heterogeneous catalysts a great advantage over homogeneous catalysts. On the other hand, homogeneous catalysts are often more active and more selective than heterogeneous ones. A great advantage would be gained if the two types could be combined with retention of the best properties of each. A very small step toward the homogenizing of heterogeneous catalysts was made by the development of fluid-bed catalysts, but, until recent years, chemists have made few attempts to heterogenize homogeneous catalysts. Most of the attempts which have been made are in biochemistry, perhaps because biochemists have known for a long time that enzymes, which are catalysts in life processes, are often in a heterogenized state because they are bound to cell membranes. Following this lead, they have adsorbed or chemically linked enzymes to glass, cellulose, and other insoluble materials, and have found that the heterocatalysts thus formed are superior to homogeneous ones in several ways. The chemical industry has only recently begun to bind homogeneous catalysts to solid supports, and a great deal of research remains to be done in this field. It is the purpose of this report to encourage such research, and to suggest some guide lines for it.

THE SUPPORT

Physical Aspects

Surface Area

For a heterogeneous catalyst to operate efficiently it is necessary for the catalytic sites to be readily accessible to reagents. This has to be combined with a reasonable number of sites per unit volume of support. To obtain this condition the support must be intrinsically porous or must become

† Although the term 'heterogenizing 'has been used here, in enzyme chemistry the term 'immobilizing' has been recommended.

porous through swelling under the conditions of the reaction. If, however the rates of the chemical reactions are fast relative to the rate of diffusion into the pores a significant number of sites may not be utilized. Thi has to be taken into account when choosing the support.

Mechanical and Thermal Properties

Because heterogeneous catalysts are subject to mechanical stresses they must be resistant to fracture and compression. On the other hand for some reactions the heat transfer from the reacting medium should be an important factor in order to avoid the formation of hot spots in the catalytic bed. Many organic polymers are subject to thermal collapse and enzymes are sensi tive to thermal deactivation. The importance of this point can be emphasiz by taking into account the low thermal conductivity of polymer-particle bea when employed as a support. This fact implies the employment of thermostab polymers. In spite of this, highly exothermic reactions will have to be pe formed in fluidized beds or they will require particular technological deve opments such as the employment of metallic supports to which the catalyst i linked through an oxide layer. Catalyst binding of this type has been achi for enzymes.

Chemical Properties

Chemical Stability

The support and the link between the active site and the support must be inert to the conditions of the catalytic reaction. They must not react wit the solvent or reagents in it. If such reaction occurs, the support will degrade to soluble species or the chemical link will be broken.

Microenvironment

In addition to acting as a physical barrier for the catalytic site, the support can have an appreciable effect on the catalytic reaction itself. This has been shown for immobilized enzymes as well as conventional heterogeneous catalysts.

pH Effects

If the support carries acidic or basic groups then the pH within the support can be different from that of the bulk solution. This is especially important in the case of enzyme catalysts which have optimal pH *v.* activity response. For any catalyst which one desires to heterogenize this effect has to be considered.

Promoters

For certain immobilized enzymes the presence of K^+ associated with adjoining carboxyl groups is believed to have a promoting effect on the activity of the catalyst. This resembles the known effect of trace metals in many commercial catalysts. To our knowledge no examples have been reported for other heterogenized catalytic systems, but they are certainly worthwhile seeking.

Specific Affinity of Support for Substrate or Solvent

A support which has greater affinity for the substrate than for the solvent will increase the concentration of substrate in the neighborhood of the catalytic site. In the area of immobilized enzymes, the favorable effect on enzyme activity of supports specifically adsorbing the enzyme substrate in question has already been demonstrated. In the experiment referred to, hydrophobic matrices have been preferred for enzyme reactions involving hydrophobic substrates. Porous hydrocarbon polymers are known to sorb oils from aqueous solutions and this principle can be utilized in catalyst preparation and use.

Steric Effects

The bulk of the support can sterically influence the course of the reaction and even prevent its occurrence. The character of the linkage between catalyst and support is critical in this respect. With immobilized enzymes its length is a determining factor, while with transition-metal complexes also the distance between the functional groups plays an important role.

These properties can be utilized to promote certain desired effects. The selectivity towards a sterically less demanding reaction can be increased or rate increases can be obtained by forcing a metal into more active coordination geometry by the bound ligands. Enzymatic reactions have their stereospecificity built in but for other systems it may be possible to introduce this into the support. An example of this is the stereospecific hydrogenation brought about by noble metals supported on natural silk.

Materials

Below is given a list of supports which have been applied for the heterogenizing of catalysts.

Inorganic supports:	glass silica clay metal zeolite oxides
Organic supports:	polyamino acids acrylic polymers styrene polymers cellulose cross-linked dextrans agarose

Of these the organic supports have received the most attention but the potentials of inorganic supports should not be overlooked.

MODES OF ATTACHMENT

Adsorption

Gas-Phase Reactions

As elution of the catalytic material by the gas stream is not likely to be a problem, conventional techniques can be utilized to deposit the catalyst on the support. A less conventional method has been to have a homogeneous catalyst dissolved in a high-boiling solvent within the pores of a support.

Liquid-Phase Reactions

If the catalyst is not to be eluted it has to be insoluble in the liquid reaction mixture, but it may be deposited from another solvent. Even if a certain degree of elution occurs the catalyst can be readsorbed on a fresh bed of support and may then be re-used. This has been demonstrated with immobilized enzymes and metal complexes.

Chemical Linkage of Catalyst to Support

In order to achieve chemical linkage between the catalyst and support, the latter must have functionality. This may be an intrinsic property of the solid (e.g. silica gel) or it can be introduced by chemical modification. In principle, the catalyst may be linked to the functional group as a unit; this has been done in the case of enzymes and has been reported for complex transition-metal catalysts. Frequently, however, the catalyst has to be built up from its component parts, e.g. starting from polystyrene, the P, butyl groups, Rh, CO and hydrogen have to be introduced step by step in making the catalyst:

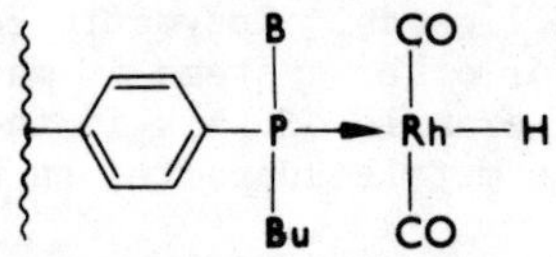

B CO
P → Rh — H
Bu CO

The functionality may serve to link the catalyst to the support in a variety of ways, some of which are illustrated below.

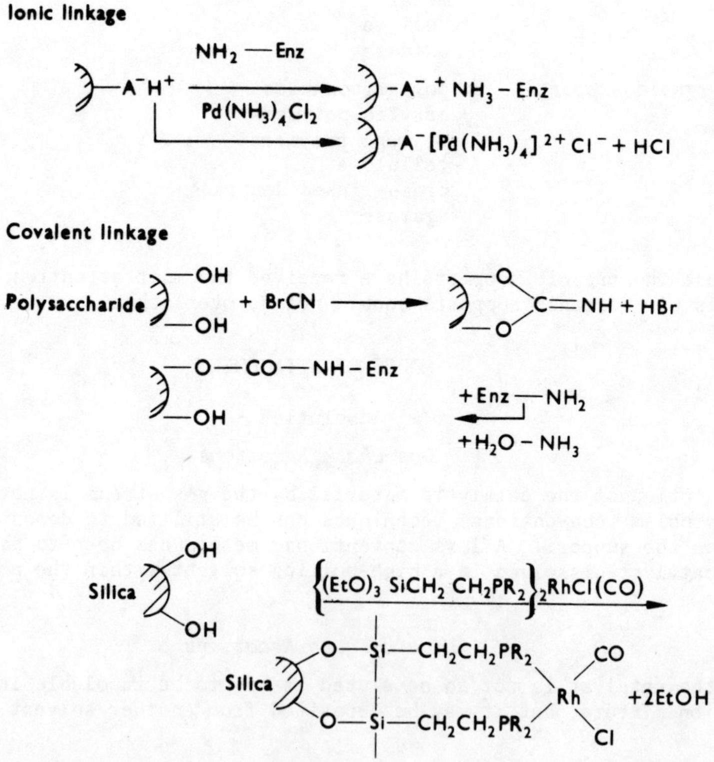

Ionic linkage

$$\text{—}A^-H^+ \xrightarrow{NH_2\text{—}Enz} \text{—}A^- \; {}^+NH_3\text{—}Enz$$

$$\text{—}A^-H^+ \xrightarrow{Pd(NH_3)_4Cl_2} \text{—}A^-[Pd(NH_3)_4]^{2+}Cl^- + HCl$$

Covalent linkage

$$\text{Polysaccharide}(\text{—OH})_2 + BrCN \longrightarrow (\text{—O})_2C{=}NH + HBr$$

$$\xrightarrow[+H_2O\,-\,NH_3]{+Enz\text{—}NH_2} \text{—O—CO—NH—Enz},\ \text{—OH}$$

$$\text{Silica}(\text{OH})_2 \xrightarrow{\{(EtO)_3SiCH_2CH_2PR_2\}_2RhCl(CO)}$$

$$\text{Silica}(\text{—O—Si—}CH_2CH_2PR_2)_2Rh(CO)(Cl) + 2EtOH$$

Aggregation

Polymerization

As an alternative to fixing the catalyst to a support, it is possible in some cases to polymerize the catalyst. In order to facilitate this, a polymerizable function has to be linked to the catalyst (e.g. an ethylene group to a phenyl nucleus). Examples of this procedure include olefin polymerization, transesterification or polycondensation. In these synthetic methods there is flexibility in the choice of the length of the linkage. By choosing the right polymerization conditions almost any physical shape of polymers can be formed.

Crosslinking

If two functional groups are attached to the catalyst, a crosslinking procedure can be used:

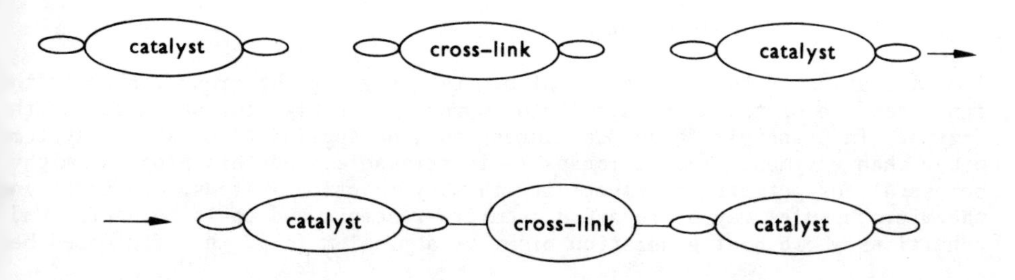

(e.g. glutaraldehyde $\mathrm{{}^{O}_{H}\!\!>C\text{-}(CH_2)_3\text{-}C\!\!<^{O}_{H}}$, which reacts with amino groups on the catalyst). Except for aggregation this method might also be used for forming a skin of catalyst-containing material around a metal or metallo-magnetic core.

Entrapment and Microencapsulation

Biological Systems

Entrapment in a Gel

Enzymes can also be immobilized by entrapment within a polymeric matrix (e.g. polyacrylamide gel)

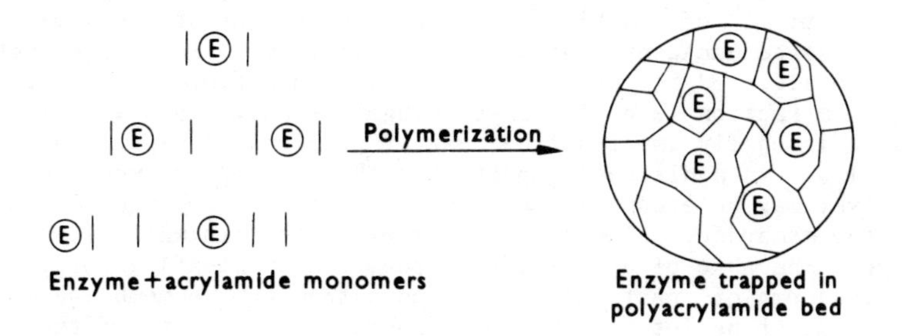

Microencapsulation

A second method which is used for entrapping enzymes is to enclose them in nylon microcapsules. The process involves the formation of an emulsion of an aqueous phase, containing the enzyme, in a non-polar solvent. The two materials which form the nylon, a diamine and a dicarboxylic acid, polymer- ize at the polar/nonpolar interfaces. The extent of emulsification contro the size of the capsule while the extent of polymerization controls the po size. Thus large or small microcapsules having either large or small pore sizes are possible. This technique of polymerization is well-known in the synthesis of porous organic polymers.

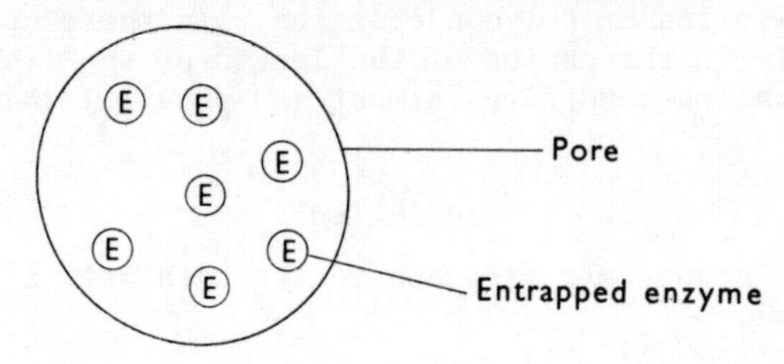

Microcapsule

A modification of this technique allows the enzyme to be cross-linked to th inner surface of the microcapsule to improve (possibly) the stability of th enzyme. In principle these procedures could be applied to catalytic system other than enzymes. The microcapsule is permeable, and this property might be useful for permitting selectivity in many reaction systems. In additio the microcapsules may be regarded as micro-reactors and could be useful und conditions where heat generation might be a problem (e.g. in a fluidized be

In Non-Biological Systems

The best non-biological analogue to the described enzymatic techniques is the established use of molecular sieves, containing small metal crystallite within their pores. The pore admits only molecules below a certain size, and these are therefore selected for reaction.

FORM OF THE CATALYST

The solid may be in the form of beads or pellets which can be used in a fixed or fluidized bed, with gaseous reactants and products. The catalyst particles may also be slurried in a liquid medium. Pellet size is importan because of pressure gradients due to diffusion and mass transport in the channels between the particles. The porosity of the bead or pellet itself is an additional significant factor in terms of accessibility to reactants, diffusion control of the latter and of the products, and consequent influen on the selectivity of reaction. Size and mechanical strength may also have to be considered because of stability and lifetime. Adverse effects could result from elevated temperatures, attrition in a fluidized system, and swel ling and contraction (e.g. in ion-exchange resins). In a slurried phase, size is additionally important because of problems in separation and regene ation cycles. An additional important technique employs membranes. Here the catalyst may be encapsulated as a free entity, or bound to the inner wall of the membrane. Fine tubular systems with thin walls seem to be most advantageous because of high surface area, reasonable flow rate, and rapid- ity of transport of reactants and products through the membrane walls. As with the bead or pellet catalysts, a compromise in length of the tubes, dia- meter, thickness of walls, and pore size may have to be reached, because of stability, mass transport, segregation of feed components, and access of particular reactants. In all cases the form of the catalyst may also have to be taken into consideration in design because of heat control.

POTENTIAL OF HETEROGENIZING CATALYSTS

Ease of Separation, Reuse and Recycle

A heterogenized catalyst allows much greater versatility in the execution of the catalytic reaction. Catalyst beads can be used in fixed beds, fluidized beds, or slurries, after which the catalyst can be recovered by filtration or sedimentation. However, the use of two immiscible liquid phases, one containing the catalyst and the other containing the substrate, must be considered. The two phases can be separated by conventional means and high degreees of dispersion can be obtained through emulsification.

This ease of separation may be particularly advantageous in situations where frequent catalyst regeneration or reactivation is required.

Engineering Flexibility

The ability to retain enzymatic and other catalytic activity in many different physical and structural forms, such as membranes, films, beads, tubes, microcapsules, etc. allows considerable flexibility in the design of reactors.

Reactor configurations which are presently in use for immoblized enzymes include hollow fibers or coiled tubes in which enzymes are enclosed or are recirculated, coiled films to increase the surface area, and enzyme-containing membranes through which the substrate is forced under pressure. The use of 'tea bags', packed-beds and fluidized beds for heterogenized catalysts in bead form (e.g. glass, organic polymer microcapsules, etc.) has been found useful for both enzyme and transition-metal catalysts. Each has its own particular advantage and many of these reactor configurations are immediately applicable to other types of heterogenized catalysts.

Stability

Enzymes, because their activity depends on a highly specific three-dimensional conformation (tertiary structure), are highly sensitive to forces which tend to destroy intramolecular bonding and hence cause them to unfold. Attachment of the enzyme molecule to a rigid matrix by more than one bond might stabilize the active, folded state. This principle may also apply to non-enzymatic catalysts which require stabilization of special structures or configurations.

Multifunctionality

Multifunctional catalysts are already known and utilized in conventional heterogeneous processes. It also appears possible to utilize similar techniques in the application of immobilized enzymes and transition-metal complexes. Many enzymatic processes are known to operate through a sequence of catalytic reactions. The importance of heterogeneous multifunctional catalysis is illustrated for a sequential reaction below:

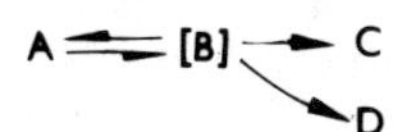

Both activity and selectivity in this example are dependent on [B]/L where L is the diffusional path of the intermediate product between catalytic sites. The shorter the distance over which the intermediate must diffuse (i.e. the greater the proximity of the sites), the higher the activity and selectivit which can be achieved. Attachment of both co-factor and enzyme or a multiplicity of enzymes within the same support matrix could have particular advantage.

Another way in which multifunctionality in a catalyst can affect the activity and selectivity is where more than one type of catalytic site can simultaneously influence the substrate. This phenomenon is well characterized for enzyme catalysts which have many functions within the active site. It appears possible to develop many applications in which advantage is taken o systems containing a multiplicity of functional sites. A particularly inte esting aspect of the incorporation of a multiplicity of sites onto a solid is that in this way it may be possible to utilize a number of catalysts whi would be incompatible in homogeneous solution.

Catalyst Concentration

From a technological point of view one very important aspect of conversio of a catalyst to a heterogeneous form is that it provides a means to carry out a given reaction in a much reduced reactor volume. This is particularly significant in reactions for which the catalyst has poor solubility in the desired reaction medium. In a sense, heterogenizing a catalyst frees it from solvent constraints. This feature may allow the use of catalysts for reactions which are not possible homogeneously or have poor selectivity in available solvents. Concentration of catalyst into a small volume can greatly reduce the cost of a reactor in which it is to be employed and allows a much higher conversion of reactants in a given volume. The effective concentration of transition-metal complexes has been improved over 100-fold. This same principle may have application with immobilized enzymes.

Corrosion Prevention

In some homogeneous catalytic systems the solution of the metal complexes may react with the reaction vessel but this may be avoided by isolation of the catalyst. For example in the Wacker and related processes for oxidation of ethylene using Pd salts, the solutions are corrosive and alloy steels are needed for the reactors. Supporting the Pd on carbon reduces this costly problem. Another aspect of this phenomenon arises when the catalyst metal is lower in the electrochemical series or forms less stable complexes than the material of the reaction vessel. Replacement can occur and the catalyst may be lost from the reaction medium by deposition on the walls. If the metal as a complex is made insoluble on a support this will not happen.

Medical Engineering

The clinical potentialities of enzyme technology are substantial. Of the more than 120 anomalies and diseases known as inborn errors of metabolism, most represent enzyme deficiency diseases in which certain enzymes normally found in the body are either lacking or inactive. A direct approach in many of these would be to supply the patient with the missing enzyme. However, the foreign protein would immediately produce an adverse immunological reaction in the patient. On entrapping the enzyme, however, in a gel, this reaction could be prevented. The enzymes could be enclosed in tiny semipermeable polymer beads using a polymer compatible with the human organism. The beads could be introduced directly into the bloodstream of the patient or alternatively packed in a shunt chamber connected to the circulatory system. A promising side-line to this approach would lie in the construction of a new type of artificial kidney. Enzymes such as urease, the latter degrading urea together with appropriate sorbents would be entrapped within polymer-beads.

These beads or microcapsules could then be packed in a chamber connected to the patient's bloodstream. Such a system would represent a valuable alternative to the bulky and costly dialysis procedure normally employed as an artificial kidney.

CONCLUSIONS AND RECOMMENDATIONS

The work which has been reported on the heterogenization of soluble catalysts is scanty, but it is sufficient to show that heterogenization can be effected in a great variety of ways. However, many problems remain to be solved.

In some cases, heterogenization produces catalysts which have desirable properties other than ease of separation of the catalysts from the substrate. This is particularly apparent with biological catalysts, but can certainly be achieved with synthetic ones as well. This is one area in which research is badly needed.

It has long been known that heterogeneous catalysts deposited on different supports may give quite different results. Apparently, there has been no study of this property in connection with heterogenized homogeneous catalysts, and research in this field is strongly urged.

The stereochemistry of heterogenized homogeneous catalysts covers a wide range, and changes in the stereochemistry may make possible a wide variety of results from the same catalyst. For example, an optically active catalyst or support may bring about reactions of only one isomer of an asymmetric substrate. Here, again, much research is needed.

For proper understanding of the reactions of metal-complex catalysts, there is still a need to determine the ligand structures in catalytically active complexes during the course of reaction. For example, it is possible to use supported complexes to throw light on the dissociation of ligands by examining the release of metals from polymers carrying different ligands.

Liganding elements already used for attaching complexes and enzymes to solids include O, N, S, P, and As, and a variety of solid supports has been used, but there is still scope to explore these areas for particular applications.

Synthetic techniques should be developed for the construction of sophisticated multifunctional heterogeneous catalysts based on metal complexes, enzymes, and models of enzymes.

We fully appreciate the efforts of enzymologists to study the structures of prosthetic groups, their isolation and the mechanisms of action of enzymes, and we support such work. Advances here will lead to new and valuable catalyst systems.

LITERATURE

WINGARD, L.B., Jr. (ed.) (1972). *Enzyme Engineering*, (Interscience, New York).

KOHLER, N. and DAWANS, F. (1972). La catalyse par des dérivés de métaux de transition déposé sur ces supports, polymériques organiques, *Rev. Inst. Franç. Pétrole, Ann. Combust. Liquides*, 27, 105.

STARK, G.R. (ed.) (1971). *Biochemical Aspects of Reactions on Solid Supports*, (Academic Press, London and New York).

CONCLUSIONS

I. GENERAL COMMENTS

Catalysis is an essential factor in chemical technology, contributing greatly to world economy, and is the basis of most life processes in biological systems. Research has played a vital part in the development of industrial processes and in the understanding of life systems. Since the future of new reactions and processes will be largely catalytic, since catalysis will be heavily involved in any energy supply system, and since catalysis will be essential in improving the quality of life, research in all areas of catalysis should be strongly encouraged.

A series of proposals for future research, first of a general nature and then related to specific reactions follows this section. It is to be emphasized that the function of the conference was to study the areas of overlap among three areas of catalysis, homogeneous, heterogeneous and metalloenzyme, and the proposals derive only from these areas of overlap. We have not studied and we present no proposals as to items primarily within one area of catalysis nor to items involving overlap of any of the areas with external areas. The order of listing is random and no order of importance is implied. The reader is referred to the working group reports for more specific conclusions within the areas covered by each group.

As the conference proceeded, it became overwhelmingly clear that the mutual interaction between scientists in the different disciplines was extremely stimulating and of great potential value for the general development of the science of catalysis.

II. GENERAL TOPICS

The following work is strongly recommended:

1) Procurement of thermodynamic data relating to the binding of hydrocarbon, hydrogen, nitrogen, oxygen and other species to active sites of catalysts. These data are needed for a more penetrating understanding of the elementary processes of the catalytic cycle.

2) Preparation and distribution of standard catalytic materials to facilitate comparison of results obtained by experimentalists in various laboratories.

3) Vigorous research on the characterization of active sites, by all applicable tools. Despite extensive investigations in this area the current level of understanding is inadequate. More extensive and more accurate estimations of the number of active sites should lead to a uniform presentation of kinetic data in terms of turnover numbers; this is highly desirable to facilitate comparison of results in the three disciplines. It would also assist in the evolution of microenvironmental factors such as solvents, promoters and neighbouring functional groups.

4) The development of new methods for the investigation of transient intermediates and the extension of existing methods (spectroscopic, isotopic and kinetic) for the identification of reaction intermediates.

5) Development of improved and extended methods for the attachment of homogeneous and enzyme catalysts to supporting matrices and the study and characterization of the resultant heterogeneous catalysts. Such systems provide not only the operational advantages of heterogeneous catalysis but also a means of altering activity and selectivity, and of making novel multifunctional catalysts.

III. SPECIFIC TOPICS

The following work is strongly recommended:

1) The extension and development of heterogeneous and homogeneous catalysts capable of bringing about stereospecific hydrogenations and oxidations.

2) Studies of methods of selectively attaching ligands to surfaces of heterogeneous catalysts, as a means of moderating the activity of such catalysts for high temperature alkane reactions.

3) The development of metal cluster compounds as homogeneous catalysts and further related study of very small metal crystallites in heterogeneous catalysis.

4) The characterization of alloy catalysts and related polynuclear complexes in homogeneous catalysis.

5) The study of model systems in homogeneous catalysis which can aid in the understanding of rate enhancement, the nature of reactive species and of product stereoselectivity in enzyme catalysis. Systems for initial study might well include those for the activation of dioxygen or dinitrogen.

6) The search for exceptionally reactive homogeneous metal ion complexes with a potential for the activation of dinitrogen and saturated hydrocarbons as possible catalysts capable of operating at moderate temperatures.

7) Studies of mechanistic features of selective oxidation and oxygenation in all three fields of catalysis. These should include studies of the geometry of initial bonding of dioxygen to complexes, metalloenzymes and surfaces. Selective oxidation is of key importance, both mechanistically and technologically.

8) Improvement in understanding of hydrogen activation by heterogeneous systems, and subsequent reactions of activated hydrogen with organic substrates, in heterogeneous and homogeneous systems. We encourage the search for enzymatic activation of hydrogen and further study of the heterogeneous and homogeneous hydrogenation of carbon dioxide, carbon monoxide and nitric oxide. This is particularly important in preparation for the possible development of a 'hydrogen fuel economy' which would greatly favour the use of hydrogenation processes.

9) Clarification of the mechanisms in known dinitrogen reduction processes by the identification and characterization of reactive intermediates and study of model systems in all three fields of catalysis. This is necessary in order to devise better catalysts for nitrogen fixation which is important for industrial and nutritional purposes.

SUBJECT INDEX

Active centers 2,5f,58f,75f,174
 changes in solvent structure at 5
 defects as 75
 in homogeneous catalysis 107, 134
 polarity of microenvironment 2
 vicinal groups at, in enzymatic catalysis 2
Adenosine diphosphate 30,43,44
Adenosine triphosphate 30,43,150
Adrenoxin 29
Alchohol hydrogenase 47
Alchohol dehydrogenase 4,10,17
 metal ion specificity 17
Aldehyde oxidase 48
Alkaline phosphatase 4,10,17f,25f
 defective in bone calcification failure 26
 from *Escherichia coli* 25
 mechanism of activity 28
 metal ion specificity 17f
 structure 26
Allosteric catalysis 2
Allosteric proteins 46
Allosteric site 2
Anatase 175
Aspartate transcarbamylase 10
Ascorbate oxidase 154
Associative desorption 56
Associative surface reaction 56
Asymmetric hydroformylation 146
Auger electron spectroscopy 78, 135,147,166
Azotobacter vinelandii 150,151
Azurins 49

Batch reactor 51
Binuclear metal oxygen complexes 158
Bonhoeffer-Farkas mechanism 55
'Burst kinetics' 26

C-C bond activation 170f
C-H bond activation 167f
Carbonic anhydrase 4,10,17f
 inhibition by sulfonamides 19
 metal ion specificity 17f
 reactions 19
 similarity to carboxypeptidase A 22
 structure 21
Carbonium ion chemistry 72f,103f, 131,135
Carboxypeptidase A 4,9f,10,22,44
 C-terminal peptide cleavage 11
 mechanism of activity 15
 metal ion specificity 17f
 structure 13
Catalase 47,49
Catecholase 160
Charge transfer complexes 143
Chlorophyll 30
Chromatium vinosum 29,32
Clostridium pasteurianum 29,34, 150,151
Circular dichroism 2
Cobalamins 6
Coenzyme B_{12} 4,49,166
Collective electrons 56f
Coordinatively unsaturated surface 56f
Corrins 118,169
CNDO 137
Cracking 166
Cubane complexes 37,166
Cyclodextran, hydrophobic cavity of 8
Cytochrome b_5 10,49
Cytochrome *c* 4,10
Cytochrome *P*-450 48,158
Cytochrome oxidase 39,47

d-d transition optical spectra 20
Dehydrocyclization 73,166
 of heptane 73
Dehydrogenation 73,141,142,162, 166,174
 of alkyclohexanes 73
Demanding reaction 66,77,135
Desaturases 162
Dielectric relaxation methods 133
Diimine complexes 107,153
Differential reactor 52
Diffusion in heterogeneous catalysts 58,182
Dinitrogen complexes 152

Dioxygenases 160
Dissociative adsorption 56,68
Dissociative surface reaction 56
Dopamine hydroxylase 159
Double resonance spectra 147
  electron-nuclear 147
  nuclear-nuclear 147
Dual functional catalysts 73,169
  dehydrogenation of alkyl cyclohexanes 73
  dehydrogenation of heptane 73
  isomerization of methylcyclopentanes 73
  isomerization of straight chain alkanes 73

Electrocatalysis 143,144,175
  synthesis by 144
Electronic fluorescence spectroscopy 133
Enolase 45
Enol stabilization 116
Epoxidation of ethylene 159
ESCA 133,135,147
Escherichia coli 25
ESR 2,20,21,32,43,133,137,146,152,157,158,164
  hyperfine structure 133
  spin labelled ligands 137

Facile reaction 61,76,106,135,
FAD 4
Fatty acid desaturase 156
Ferredoxins 4,10,28f,29,49,150
  functions 30
    in anaerobic bacteria 30
    in green plants 30
    photosynthetic bacteria 32
    synthesis of acetyl coenzyme A 31
  oxidation-reduction potentials 29
  sources 29
  structure 32
    high potential iron protein 35
    Mircrococcus aerogenes ferredoxin 36
    rubredoxin 34
Ferroheme 160
  O_2-complex 160
Field emission microscopy 153
Flash desorption techniques 135,153
Flash filament 133
Flavins 145,162
Flavin requiring enzymes 145
Flow reactor 51
Fluidised bed catalysts 177,183
Fructose 1,6 diphosphatase 46
Fuel cells 141,144,175

Galactose oxidase 47
Gas-adsorbate reaction 56

Haber catalyst 153
Hard acid 57,108
Hard base 57,108
Heme 1,4,156,160
Heme proteins 29
Hemocyanins 49
Hemoglobin 27,39,49,158,160
  example of positive cooperativity 27
  high spin-low spin change 39
Heterogeneous catalysis, control of 173
Heterogeneous reaction classification 56
Heterolytic splitting of H_2 111,122,177
High potential iron protein 10,29,32f
  structure 35
Histidine deaminase 44,45
Homogeneous activation of saturated hydrocarbons 128
Horiuti-Polanyi mechanism 66
Human carbonic anhydrase *C* 10,22f
  structural water 23
  structure 22
Hydrazine complexes 153
Hydride ion transfer 1,145
Hydrocarbon skeletal rearrangement 166,170
Hydrocyclization of paraffins 135
Hydrogenation 52,128,141f
  double bond isomerization 143
  geometrical isomerization 143
  hydrogen activation without added substrate 142
  hydrogen transfer 142
  hydrogenolysis 143
  via metal hydrides 145
  oxidative dehydrogenation 142
  reductive hydrogenation of inorganic compounds 143,145
  skeletal isomerization of hydrocarbons 143
  of unsaturated functions 142

Hydrogen fixing bacteria 145
Hydrogenolysis 143,160,170
Hydroisomerization of paraffins
 135
Hydrophobic cavity of cyclodextran
 8
Hydrophobic matrices 179
Hydrophobic substrates 179
Hydroxylanes 162
Hydroxylation 161
p-hydroxyphenylpyruvate oxygenase
 48,49

Induced enzyme synthesis 139
Inhibition of heterogeneous
 catalysis 86
Insertion reactions 72,110
Ion scattering spectroscopy 135
IR emission spectroscopy 133
IR spectra 32,77,79,133,164,165
 forbidden transitions induced
 by electric fields 81
 of surface bound groups 164
Isolation matrix technique 134
Isomerization of *n*-butenes 103f
 1-butene isomerization 104
Isomerization of methylcyclo-
 pentanes 73

Jahn-Teller effect 18

Kinases 43f
Kinetics of heterogeneous reactions
 58f
 dual functional catalysts 73
 hydrogenation of acetylenes 64f
 hydrogenation of cycloalkanes
 69f
 hydrogenation of dienes 66f
 hydrogenation of nitrogen 63f
 hydrogenation of olefins 68f
 reactions on oxides 71f
 reactions on silica-aluminas
 72f
Klebsiella pneumonia 150,151

Laccase 156
Langmuir adsorption isotherm 1,
 2,59,61,62
 saturation 1,2
Langmuir equation 132
Langmuir-Hinshelwood mechanism
 62,131
Laser-Raman spectroscopy 133
Low energy electron diffraction
 61,78,147,153,166
Lindlar's catalyst 65,173
Linear polyethylene 52
Lipoxygenases 161

Mass spectroscopy 65
Mass transport in heterogeneous
 catalysis 182
Medical engineering in enzyme
 deficiency 184
Metal atom clusters 136
Metallic oxide catalysts 91
 complete oxidation reactions
 on 91f
 deoxidation reactions 98
 N_2O and NO decomposition on
 98
 isotopic oxygen exchange on
 93f
 partial oxidation reactions
 100
 acrolein synthesis 100
 partial oxidation of
 methanol 102
 2-propanol decomposition 101
Metal complexes in homogeneous
 catalysis 107
 aggregation 136
 chemistry 136
 coordination numbers 112,136
 solvation numbers 136
 stabilization or reactive
 species 107
 substrates as potential
 ligands 107
Metal ion specificity of enzymes
 17
 Jahn-Teller effect for Cu(II)
 18
Methylmalonyl coenzyme A isomerase 4
Michaelis complex 4
Michaelis constant 1
Michaelis-Menten equation 132
Micrococcus aerogenes 10,29,34,36
Microwave spectroscopy 133,139
Mitochondria 9,139
 respiratory chain 9
 role of 139
Molecular hydrogen activation 128
Molten salt catalysts 173
Mono-oxygenase 161,162
Mono-oxygenation 161
Mössbauer spectroscopy 2,33,36,
 38,133

Multifunctional catalyst 183
Multinuclear metal complex 153
Muscle pyruvate kinase 44
Mutated nitrogen-fixing
bacteria 152,153
Myoglobin 49

Nitrogenase 4,32,49,151
Nitrogen fixation 128
NMR 2,34,44,45,65,133,136,146
broad-line NMR 133
Fourier transform NMR 134
single pulse Fourier
transform NMR 137
high resolution NMR 133,147
pulsed NMR 133,139
Non-heme Fe proteins 29

Olefin dismutation 52,127
Olefin metathesis 108
Olefin oxidation over noble metal
catalysts 104
acrolein formation 105
acrolein partial oxidation of
106
2-butenes 105
ethylene 104
propylene 105
Olefin reactions under homo-
geneous catalysis 108
Optically active catalyst 185
Optically active holes 146
Overhauser effects 147
Oxalacetate decarboxylase 45
Oxene hypothesis 158
Oxenoid mechanism 162
Oxidative addition 128
Oxidative dehydrogenation 166,
174
Oxidative splitting of double
bonds 160
Oxidoreductases not involving
coenzyme 3
Oxygen activated dehydrogenation
of aliphatic compounds 162
Oxygen species 157f
Oxyhemoglobin 160

Packed-bed catalysis 183
Para-hydrogen conversion 80,145
Pearson's SHAB concept 57
in homogeneous catalysis 108
predicting poisoning 58
Peroxidase 1,49,160
Peroxo complexes 159
binuclear 159
mononuclear 159
Perturbation Hamiltonian 134
Phosphoenolpyruvate carboxykinase
44
Phosphotransferase 43
Photoactivation 175
of alkane reactions 175
over anatase 175
Phthalocyanins 118
Platinum catalysts 76
facile reactions over 76
reactivities of surface
atoms 76
Poisoning of heterogeneous
catalysts 86
selective poisoning 87
Polynuclear dinitrogen complexes
153
Porphyrins 118,143,169
Prostaglandin synthesase 161
Pulse radiolysis 157,163
Pyridine nucleotides 145
Pyrocatechase 48
Pyruvate dehydrogenase 31

Reforming 166
Relaxation kinetics 43,134
Reactions related to homogeneous
catalysis 109
insertion reactions 110
oxidation-reduction 109
oxidative addition 110
reductive elimination 110
substitution of ligands 109
Reactive adsorption 56
Reactive desorption 56
Reduction of dinitrogen like
compounds 153
Rideal mechanism 55
Role of metal ions in
catalysis 112f
electron transfer reactions
118f
electrophilic reactions 116f
enol stabilization by proton
loss 116f
heterolytic splitting of H_2 116
ligand stabilization 115f
nucleophilic reactions 112f
with chelation 114
with π-bonding ligands 115
with soft bases 113
oxidative addition reactions 121f

promotion of ligand synthesis 117f
symmetry restricted hydrocarbon rearrangements 125
the 'template effect' 115f,118
synthesis of macrocyclic ligands 118f
Rubredoxin 10,29,32,34
structure 34

Schrödinger equation 134
time dependent 139
Secondary ion mass spectroscopy 135
Selectivity in heterogeneous catalysis 85f
by poisoning 87
self poisoning 88
Selectivity in homogeneous catalysis 126f
by ligand change 136
by solvent change 136
Selective poisoning 87,135
Shape-selective catalysis 86
Silica-aluminas 52,72
Sodium pyruvate 150
Soft acid 57,108
Soft base 57,108
Spin conservation 155
Static reactor 51
Stereoselective synthesis 146
Stereospecificity in homogeneous catalysis 126
protective groups 127
Stopped-flow techniques 137
Stopped-flow kinetics 138
Structure insensitive reaction 61, 76,106
Structure sensitive reaction 61,77
Sulfite oxidase 47
Superoxide dismutase 4,47
Supported catalysts 58
diffusion in 58
Surface coordination numbers 75
Surface diffusion coefficient 134, 135
Surface polynuclear centers 174
Surface organometallics 54
Symmetry restricted hydrocarbon rearrangements 125

'Tea-bag' catalysts 183
Temkin-Pyzhev model 64
Temperature programmed desorption 133
Thermolysin 10
Thiamin pyrophosphate 31
T-jump kinetics 138
Turnover numbers 138,140
Trans effect ligands 108
Trypsin 11
Tryptophan pyrrolase 48,49,160
Turnover numbers 12,25,135
Tyrosinase 4,48,159

Urease 184
UV spectroscopy 133

Vitamin B_{12} 6,169

Wacker process 105,184
Work function 153

Xanthine oxidase 4,48,49
X-ray crystallography 1,9,20,43, 138,146,152
X-ray fluorescence spectroscopy 133
X-ray photoelectron spectroscopy 166
D-xylose isomerase 45

Zeolites 52,72,86,134,139,166, 168,173
constitutional protons in 134
Ziegler-Natta catalysts 108
Zinc oxide catalyst 78f